Modelling Language

Natural Language Processing (NLP)

The scope of NLP ranges from theoretical Computational Linguistics topics to highly practical Language Technology topics. The focus of the series is on new results in NLP and modern alternative theories and methodologies.

For an overview of all books published in this series, please see
http://benjamins.com/catalog/nlp

Editor

Ruslan Mitkov
University of Wolverhampton

Volume 10

Modelling Language
by Sylviane Cardey

Modelling Language

Sylviane Cardey

Institut universitaire de France and Université de Franche-Comté

John Benjamins Publishing Company

Amsterdam / Philadelphia

The paper used in this publication meets the minimum requirements of the American National Standard for Information Sciences – Permanence of Paper for Printed Library Materials, ANSI z39.48-1984.

Library of Congress Cataloging-in-Publication Data

Cardey, Sylviane.
 Modelling language / Sylviane Cardey.
 p. cm. (Natural Language Processing, ISSN 1567-8202 ; v. 10)
 Includes bibliographical references and index.
 1. Communication models. 2. Semiotics. 3. Interlanguage (Language learning)
       4. Computational linguistics. 5. System theory. 6. Natural language processing
       (Computer science) I. Title.
P99.4.M63C37   2013
006.35--dc23   2013000369
ISBN 978 90 272 4996 8 (Hb ; alk. paper)
ISBN 978 90 272 7208 9 (Eb)

John Benjamins Publishing Co. · P.O. Box 36224 · 1020 ME Amsterdam · The Netherlands
John Benjamins North America · P.O. Box 27519 · Philadelphia PA 19118-0519 · USA

Table of contents

Preface

The model presented in this book represents many years of research and direction of research.

I wish to thank Yves Gentilhomme who taught me micro-systemics. I decided to use the notion of micro-system for my research and build micro-systemic linguistics, which I based on logic, sets, partitions and relations, as a theory to apply to language analysis and generation. Realising that representing languages globally was a Utopia, but a nice dream, I thought that starting from what we wanted to demonstrate or solve might be a better way forward.

What does starting from a goal mean? It means finding the necessary phenomena and elements needed to solve the given problem. We do not need to refer individually, for example, to all syntactic phenomena to make the agreement of the past participle in French, but only to the part of syntax to which the problem is related, and likewise only the part of morphology and lexis, including semantics, concerned with the resolution of this problem.

The theory also proposes working in intension instead of extension which means working at a high level of abstraction, partitioning the elements of the language according to where they belong, whether syntax, morphology or lexis, but more often to all these phenomena at the same time. Having gathered the data and classified it, a micro-algorithmic system will then define the processing in such a way that not only will it lead to the problem's resolution, or represent what we would like to demonstrate, but it will also give us traceability which is in any case needed for scientific justification but mandatory for safety/security critical domains. A good example is our Classificatim sense mining system.

To avoid the usual question concerning a theory "Could you give an example?" the book is indeed furnished with many examples. As a result of the Erasmus Mundus programme, I have had the great fortune to have in my courses excellent students, selected after a severe admissions process, of very many nationalities, and who were curious as all good scientists are.

During my academic career I have supervised over forty PhD students, and continue to do so. These students have come from all over the world, Asia, Europe including Russia, North, Central and South America and the Middle East, and with backgrounds predominantly in linguistics, mathematics and computer science. Research is never accomplished alone, and I am indebted to them for the majority of examples concerning the many and diverse languages that they have analysed using the theory during their research. They will recognise their contributions in the book. I wish to acknowledge my gratitude to all my colleagues who have acted as external PhD examiners, and also those who have participated in several major projects based on the theory. A great thanks too to all my colleagues who have organised and chaired the XTAL conferences that I inaugurated hoping to bring together linguists, mathematicians and computer scientists who often have difficulty in listening to each other. The conferences have taken place in France, Italy, Portugal, Spain, Finland, Sweden, Iceland and Japan. I thank too my industrial colleagues, Airbus Operations SAS, who have aided me in understanding what are real applications in safety/security critical domains. Words cannot say how much I am indebted to my husband, Peter Greenfield, colleague and companion in research.

I would like to thank Professor Ruslan Mitkov, the editor of this series, along with the anonymous reviewers of the original proposal, and Kees Vaes my editor at John Benjamins.

I thank the IUF (Institut universitaire de France) which has enabled me to dedicate myself completely to research for the last four years, and for the year to come.

This book can be used by students as well as academics and industrial researchers (linguists, logicians, mathematicians, computer scientists, software engineers, quality engineers, etc.) looking for new methodologies not only for natural language processing, but wherever language and quality meet, for safety critical applications whether involving professionals, the general public or both.

Prologue

"Man is not merely *homo loquens*; he is *homo grammaticus*" (Palmer 1975: 8). The concept of language is represented through diverse languages which need to be situated in the flow of time. We can look at these languages as systems composed themselves of systems and micro-systems from the point of view of the whole or from the point of view of the components knowing that all the parts are interrelated as the stars in the galaxy and the macrocosm.

As Pascal said:

> But the parts of the world are all so related and linked to one another, that I believe it impossible to know one without the other and without the whole...Since everything then is cause and effect, dependent and supporting, mediate and immediate, and all is held together by a natural though imperceptible chain, which binds together things most distant and most different, I hold it equally impossible to know the parts without knowing the whole, and to know the whole without knowing the parts in detail. (Pascal, Pensées, English translation 1958: 12)
>
> Les parties du monde ont toutes un tel rapport et un tel enchaînement l'une avec l'autre, que je crois impossible de connoître l'une sans l'autre, et sans le tout... Donc toute chose étant causée et causante, aidée et aidante, médiatement et immédiatement, et toutes s'entretenant par un lien naturel et insensible, qui lie les plus éloignées et les plus différentes, je tiens impossible de connoître les parties sans connoître le tout, non plus que de connoître le tout sans connoître en détail les parties. (Pascal 1670: vol 4. 111–112)

Theory, methodologies and applications will meet here. From macro to micro or from micro to macro, which to choose, and how to represent what we call systems and micro-systems, components and the whole in the context of language and languages? We will start with our galaxy or languages as macro-systems and see why we need a micro-systemic approach.

Introduction

This book presents a way of considering language and languages in a similar way to observing other natural phenomena such as planets and the universe (or, rather more modestly, galaxies). Two approaches are possible, either one tries to analyse the phenomenon as a whole, or one tries to delimit some particular object, say a planet as a microcosm, in order to observe it in the context of the whole as a macrocosm, which in our example is the universe. In like manner to the telescope for observation in astronomy, one uses the microscope in biology. Whether it be the telescope or the microscope, we observe that both are based on the same science, that of optics.

As when we view the universe through a telescope, we think not only that language is too 'dim' and furthermore complex to be analysed in its entirety, but also that modelling the way the parts composing a language are interrelated is in reality very difficult.

Contrary to the current state of the art, in this book we present a way to look at language(s) in a microscopic manner which then leads to the macrocosm. As these microscopic parts are interrelated, we have to forget for the moment the traditional division into lexis, morphology, syntax and so on. Each of the elements of these micro-parts could in fact belong to different micro-parts, let us say rather micro–systems, whatever we want to demonstrate, to compose, to analyse or to generate.

We present an original way of decomposing a language and languages in order to bring into evidence norms, whether intra-language or inter-language. The notion of norm will serve as the basis for showing how a language(s) can be modelled by macro- and micro-structures to apprehend it(them) better. Our point of view, which combines linguistics and modelling by means of norms, results in a theory that is exploitable for very varied applications such as natural language processing and controlled languages which latter have the particularity of necessitating very high levels of reliability, and indeed for language applications in general where reliability is mandatory.

PART 1

System, language and its components

In this first part, System, language and its components, we firstly introduce the notion of systemicity independent of the discipline studied. We then see what linguists and grammarians have written about language as a system at different epochs. Keeping in mind that we will need to be able to describe our system of language and what it is composed of, and knowing that subsequently we will be talking of a system of systems, we are confronted with the difficulty of delimiting clearly the domains making up language. To this end we look at and review, adding as necessary our own observations, how grammarians and linguists have addressed this problem in terms of the way languages function. We address in particular language typology, lexicology, morphology and syntax, the various micro-components including the word, morphemes and syllables, parts of speech, semantics, and finally and of methodological importance, norm in language in respect of variously synchrony and diachrony, and good usage.

CHAPTER 1.1

The concept of system

1.1.1   System

Condillac (Condillac 1771: 1) wrote *"un système n'est autre chose que la disposi-*
tion des différentes parties d'un art ou d'une science dans un ordre où elles se
soutiennent toutes mutuellement" – a system is nothing other than the disposition
of the different parts of some art or science in an order where all these support
each other mutually.

   The notion of 'system' has resulted in much scientific investigation, and, depending on the disciplines involved and the authors' ideological leanings, the
concept has received various interpretations. However it would seem that these
interpretations share a certain common foundation to which it is interesting to
draw attention. Over and beyond each school's own specificities, we draw up a list
of the 'presumptive properties' of what it means to be a system as so expressed in a
large number of scientific publications. This enables seeing, in the chapters that
follow, in what manner a language and indeed languages are systems. These systems need to be brought to light by means of the images, which are often deformed,
being provided by the 'telescope' (macro approach), but firstly after having been
studied with the 'microscope' (micro approach).

   It goes without saying that our presumptions are great in number and that
such an initiative results only in an approximation which is evasive and which
needs to be contently reviewed containing as it does multiple points of view many
of which being disparate are even irreconcilable. In any case who can boast that
they have examined all of these points of view?

1.1.2   Systemicity

For these reasons and without claiming to elicit some strict but Utopian common
denominator, Yves Gentilhomme (1985: 35–36) sought to establish an inventory
of presumptive properties of systemiticity which seem to manifest themselves in
the scientific literature and not restricted just to linguistics, over and above those
proper to each school or discipline. We provide in what follows a translation of this
inventory:

There is a presumption of systemicity if and only if the object being studied possesses the following five macro-properties:

1.1.2.1   Identification: the object being studied forms a whole which is identifiable and can be isolated.

   1.1.2.1.1   It ought to be able to be grasped by means of one and only one global idea;

   1.1.2.1.2   we ought to be able to name it;

   1.1.2.1.3   we ought to be able to distinguish it from another system;

   1.1.2.1.4   we ought to be able to isolate it either materially or conceptually from its environment, whatever this be.

   This last property means that there exists a real or imaginary frontier which is imposed or deliberately constructed, impermeable or permeable, well defined or vague, which enables deciding, at least in a sufficient number of cases, what reasonably belongs or not to the system. Moreover, having said that the frontier can be more or less permeable, this means that some interaction can exist between the exterior environment and the presumed system.

1.1.2.2   Structure, elementarisation: the object being studied possessing a coherent internal organisation, it is convenient to distinguish the different parts.

   The supposed system which constitutes the object being studied can be decomposed in at least one way into more elementary components, that is to say, smaller, even more simple and sufficiently stable so as to be identified and inventoriable as sets whether defined or vague. These components maintain between themselves multiple connexions which can be coherently described notably in respect of relations and operations. Amongst these components, certain are declared to be primary, that is to say that in the framework of the analysis that has been undertaken we do not attempt to reduce them to more elementary components, whilst others appear as aggregates of primary components.

1.1.2.3   Interdependence, functionality: a subtle interrelation involving action and retroaction is enacted between the components and the links that they maintain, the existence (involving definition and determination) of the primary components being dependent on the secondary components as if these latter had for function establishing the existence of the former and vice-versa. Thus for some observer outside the system, it appears that this latter is finalised. The system constructs itself progressively as we observe it. The system can be called into question, reorganise itself completely, or cease to be considered as a system.

1.1.2.4   Originality: the whole does not reduce itself to an amorphous set of parts.

1.1.2.5   Persistence: the system ought to be able to be identified, conceived, distinguished, and named for a certain duration during which it possesses its own coherence and a particular organisation.

Language as a system

Let us now see what linguists and grammarians have written about language as a system at different epochs.

In fact we already find in the work of Antoine Meillet (1866–1936) the following citation (Guillaume 1973: 222) "*Chaque langue forme un système où tout se tient et a un plan d'une merveilleuse rigueur*"; that is "Each language forms a system where everything holds together and which has a marvellously rigorous plan".

However well before, during the XVI century, variously Scaliger (1540–1609), Sanctius (1587) and Vives (1531) were attempting to determine language's systems in a rigorous manner. Of grammar, Vives for example wrote that "it shows what is said and this according to what system" (Sanctius 1587: 13).

In France, the ideas become more refined (Trevoux 1734) during the period from the Grammaire de Port-Royal (Arnauld & Lancelot 1660) to the Encyclopédie (Diderot and d'Alembert 1751-1772). At this epoch one thought that each language is comprehended as a particular actualisation of a set of general rules.

The concept of a system "*un système, pour être suivi, doit partir d'une même imagination*" that is "for a system to be followed, it ought to start from one and the same idea" applied by Buffier (1709: 6–7) to the theory was quite naturally transferred to the language itself. It is in this sense that in 1747 Abbé Girard (Girard 1747: 2) wrote "*Les langues ont beau se former sans système et sans délibération, elles n'en sont pas moins systémiques ni moins fondées en raison*", that is "Even if languages have formed themselves with neither a system nor deliberation, they are not less systemic nor founded on reason because of this". James Harris (1751: 349) uses 'system' when he defines language thus: "A System of articulate Voices, the Symbols of our Ideas, but of those principally, which are general or universal".

Saussure (1857–1913), the father of modern linguistics, shows that language at every instant in its existence ought to present itself as an organisation. Being inherent to every language, Saussure names this organisation 'system' (his successors often speak of 'structure'). When followers of Saussure speak of 'language system' or 'language structure' they mean that linguistic elements have no reality independent of their relation to the whole.

Viggo Brøndal, in 1928 (Brøndal 1943) defines language as a system of signs: "*Dans un état de langue donné, tout est systématique; une langue quelconque est constituée par des ensembles où tout se tient: système des sons (ou phonèmes), système des formes ou des mots (morphème et sémantèmes). Qui dit système, dit ensemble coherent; si tout se tient, chaque terme doit dépendre de l'autre"*; that is "In a state of some given language, everything is systematic; a language, whatever it be, is made up of sets where everything holds together: systems of sounds (or phonemes), systems of forms or words (morphemes and semantemes). When we speak of system we mean coherent whole; if everything holds together, each and every term ought to depend on the others".

For Chomsky (1957), grammar, which can be considered as a total description of some language, is just a formal system where the mechanical application of rules produces the acceptable utterances. For the logician, *"le mot formel ne signifie rien d'autre que logiquement ordonné, univoque et absolument explicite"* (Gladkij 1969); that is "the word formal signifies nothing other than logically ordered, univoke and absolutely explicit".

Finally we cite Whitney (1867: 91) where the following crucial words appear: "A spoken alphabet [...] is no chaos, but an orderly system of articulations, with ties of relationship running through it in every direction.", and in the following citation (Whitney 1867: 50) "A Language is, in very truth, a grand system, of a highly complicated and symmetrical structure; it is fully comparable with an organized body." The linguist's task is defined there as "If we were to consider all these definitions of language, we can think that grammar and language are not just some indescribable chaos but rather a system" (Whitney 1875: 57).

Contrary to the mathematical concept of set, in set language in respect of a system we include simultaneously as constituents the set of elements and the set of relations that can become established between the considered elements (Gentilhomme 1985).

Let us now see how to describe our system and what it is composed of, knowing that subsequently we will be talking of a system of systems.

1.2.1   Grammatical system

For a long-time one has talked about the grammatical system.

The first problem that we are confronted with is that of knowing what the term grammar represents and what it means.

Authors who have written on the subject do not seem to agree very well concerning the terms designating the different parts of grammar. Over the last century, the metalanguage developed for studying language has become considerably enriched. To the very ancient grammatical terms etymology and syntax have been added lexicology, linguistics, phonetics, morphology, phonology, stylistics, semantics and many more too. To grammar as a discipline various sciences have been added that enable studying language (phonetics, phonology, lexicology, semantics, morphology, syntax, etc.).

Let us concentrate in particular on the domains of lexicology, morphology, syntax and semantics. However, as we will see, it is difficult to eliminate totally the domains of phonetics and phonology when we study the above domains. In fact all these domains form a system in which the different parts are very difficult to delimit.

Why is it difficult to delimit clearly these domains? Let us briefly look at how languages function.

1.2.2   Language typology

We find nothing before the XIX century comparable for language with what Aristotle had done for the elucidation of the natural sciences. As for language we see what is comparable for the first time with the work of the Schlegel brothers (1808 and 1818) which involves a classification of languages based on discriminants: (1) languages without combinations of forms, (2) languages with affixes, and (3) inflexional languages (of which the last are the most perfect according to A.-W Schlegel). The oldest languages would thus be the monosyllabic languages as for example Chinese. Those of the others which are radical-affix languages enclose a complete sense when the radical and affix(es) are associated. Amongst the inflexional languages, there would be those said to be synthetic (with case endings and with neither articles nor auxiliaries) and those said to be analytic, issue from the former. Humboldt (1836), Schleicher (1866), F. Bopp (1833) and A.-F. Pott (1849) continued with this work. Pott, basing his study on a distinction between matter (root with the main sense) and form (derivations and secondary senses), divided languages as independent or isolating such as Chinese, agglutinative such as Turkish, inflexional such as the Indo-European languages, and incorporating such as Eskimo. H. Steinthal (1860) and his pupil F. Misteli kept the preceding classifications, and Misteli added the coordinating discriminant. Nowadays, the majority of linguists are in agreement with the classification: inflexional, agglutinative and isolating. This classification can be criticised on the grounds that it only takes account of morphology. As a result Sapir gave the criterion of word construction only a secondary role (Sapir 1921). If all languages express concrete concepts designating objects, qualities or actions (they are expressed by the nominal and verbal radicals in the Indo-European languages) and also abstract relational concepts establishing the principal syntactic relations, certain languages have variously neither derivational concepts that modify the sense of concrete concepts expressed for example by diminutives, prefixes and suffixes, nor concrete relational concepts such as number and gender. According to whether a given language expresses neither, one, or the other or indeed both of these notional concepts, we should be able to group languages in classes which, depending on the nature of the criteria used, would not necessarily have a genetic character. A more recent attempt is that of Greenberg (1966) which is based on the order of words in a proposition. One distinguishes thus language types such as S(subject)-V(verb)-O(object), or S-O-V, V-S-O or even O-V-S.

What is interesting to note is that, given the reciprocal solidarity of grammatical elements (such as tense, case, person) in respect of lexical elements, we cannot borrow such a grammatical element alone in isolation but only in fact as a complete system thus rendering such an exercise not very plausible given the disruption that would occur. However it has been said that in pre-historic times language was not a means but an end; the human mind fashioned language as a work of art where it sought to represent itself. During this epoch, which has disappeared for ever, the history of language is one of creation; yet it is only by deductive methods that we can imagine the various steps that occurred. For Schleicher, for example, languages ought to have taken successively the three principal forms that appear in the classification of present day languages based on their internal structure. Firstly they were all isolating (the 'words' are units that cannot be analysed where we cannot even distinguish between a radical and grammatical elements; this is how Chinese was represented in the XIX century). Then certain languages have become agglutinative (having words with radicals and grammatical elements but without precise rules for word formation; American Indian languages are current survivors of this state). Finally, amongst the agglutinative languages, inflexional languages have developed where precise morphological rules control the word's internal organisation; these are essentially the Indo-European languages.

To conclude we cite Hagège (1986: 8) "*En réalité les langues sont des complexes de structures évolutives [...] et elles accusent très normalement des traits qui relèvent de plus d'un type à la fois*", that is "In reality languages are complexes of evolving structures [...] and it is quite normal that they show features which belong to more than one type at the same time".

Apart from language typology, a question which has been asked for a long time is "how is it possible that a sequence of separate words can represent a thought of which the primary characteristic is 'indivisibility'?" (a term used by Beauzée (1767)). Does not the fragmentation imposed by the physical nature of that which represents contradict the essential unity of that which is represented? In answering this question (the same one which in the XIX century guided Humboldt's thinkings about expressing what is a relation) general grammars state that every thought is a manifestation of the mind. Yet philosophers know how to analyse a thought in a way that, although this involves its decomposition, its unity is still respected. This is what Descartes (1637) said, (we summarise) for whom a thought comprises two faculties whose distinction does not depend on substance because their definitions are necessarily inter-dependent with each other: understanding generates ideas which are like images of things whilst the will takes decisions concerning the ideas (it affirms, denies, believes, doubts, fears, etc.). If our different thoughts also possess this structure linked to thought in general, their representation by sentences can respect their unity: it is necessary and sufficient, for this to be so, that the organisation of the words in the sentence mirror the categories as well as the relations exposed between the categories in the thought's analysis, this latter sometimes being called 'logic', sometimes 'grammatical metaphysics'. It is because of this that *"l'art d'analyser la pensée est le premier fondement de l'art de parler, ou, en d'autre termes, qu'une saine logique est le fondement de l'art de la grammaire"* (Beauzée 1767), that is "the art of analysing thought is the primary foundation of the art of speaking, or, in other terms, sound logic is the foundation of the art of grammar".

For us, this classification matters little; our model ought to be able to analyse and represent languages in general and for this it must be flexible, supple without any rigidity, and with no preconceptions.

1.2.3   Lexicology, morphology and syntax

Let us return to our four domains, those of lexicology, morphology, syntax and semantics, and let us see what grammarians and linguists have written concerning this particular division into these four domains.

For us, semantics is not a domain; rather it is what we seek to produce or recognise. We cannot put it at the same level as the other three.

For describing a language, tradition distinguishes:

1. the material means for expression (pronunciation, writing);
2. the grammar which comprises the morphology and the syntax;
3. the dictionary or lexis which indicates the words' senses.

In the Cours de linguistique générale, F. de Saussure criticises the traditional divisions of grammar which he explains thus (Saussure 1922: 133–134, English translation: 185–186):

> Grammar studies the language as a system of means of expression.
> [...]
> Grammar is conventionally restricted to *morphology* and *syntax* combined; and *lexicology* or the science of words is excluded.
> [...]

Morphology deals with the various classes of words (verbs, nouns, adjectives, pronouns, etc.) and with the different forms of flexion (conjugations, declensions). To distinguish this from syntax, it is claimed that syntax examines the functions associated with linguistic units, while morphology is merely concerned with their forms.

[...]

Forms and functions are interdependent. It is difficult if not impossible to separate them. Linguistically, morphology has no real, independent object of study: it cannot constitute a discipline distinct from syntax.

For example, in French, the personal pronoun varies in form according to its function (*il, lui*).

It seems to us that it is more appropriate to speak of morphosyntax.

In this respect Zellig S. Harris wrote in his Papers in structural and transformational linguistics (Harris 1970: 69) "The syntax and most of the morphology of a language is discovered by seeing how the morphemes occur in respect to each other in sentences".

On the other hand, is it logical to exclude lexis from grammar? At first sight, words as they are recorded in dictionaries do not seem to give rise to grammatical study which latter is usually limited to the links existing between lexical units. However we rapidly notice that a large number of such links can just as well be expressed by grammatical means. [...] Thus from a functional point of view, lexicological and syntactic devices overlap. (Saussure 1922: 187, English translation: 134).

Thus in French, *donation, donateur, donataire* have as variable elements respectively *-ation, -ateur, -ataire*, which provide signifieds which traditionally come from the lexis rather than the grammar. However transformationalists take suffixes for morphemes.

– the suffix *-ateur* in *donateur* expresses the agent as the subject of the sentence which engenders the word *donateur* (someone who gives).

– the suffix *-aire* in *donataire* expresses the beneficiary of the action as the 'attribution complement' in the sentence which engenders it.

We can admit that *carnation* agrees grammatically with the element *-carn* which is a combinatorial variant of *chair,* but the element *-vore* in *carnivore* provides a lexical signified because the action of *manger* (eat) is not implied by any of the sentence structure's syntactic functions. Elements of sense which are indisputably lexical are also expressed in French by means of prefixes and by compositional elements: the difference between *méconnaître* and *reconnaître,* between *anthologie* and *graphologie* is lexical. (Translated from Grand Larousse de la langue française 1977: article grammaire).

In French one finds for example *profond* and *peu profond* translated into English by respectively *deep* and *shallow.* We also find in English *a might-have-been* translated by *un raté* in French.

The opposition is rendered "grammatically in the first case and lexicologically in the second. [...] Moreover, any word which is not a single, unanalysable unit is essentially no different from a phrase, syntactically speaking." (Saussure 1922: 187, English translation: 134).

Thus the word *revenez* can be analysed into a prefix *re-*, a radical *-ven-* and an ending *-ez*, where the respective positions are as characteristic as the order of the subject, verb and object in the sentence. We provide in what follows a translation of what L. Guilbert (1967: 305) in his chapter on syntagmatic derivation wrote:

> Syntagmatic units are constituted from constituent elements according to a particular joining process. In the syntactic order the first of these represents the determined and plays the basic role in the construction and the second are the determiners whose determination functions according to particular syntactic models.

The system's micro-components

1.3.1 The word

Let us now examine the concept 'word', word that we have been using up to this point. Grammar books often make no attempt to give a definition of the word 'word' though they happily define other grammatical elements in terms of it. The sentence, for instance, is 'a combination of words' and the parts of speech are 'classes of words'. What a word is and how it can be defined is often not considered. The chief reason for this is that in some written languages (at least for English, French, etc.) there is no doubt at all about the status of the word. Words are clearly identified by the spaces between them. However there is quite good evidence that the word is not a natural linguistic entity. We have only to look at ancient inscriptions to see this. The use of space to indicate word division belongs to Roman times; the Greeks did not use spaces, but ran all their words together. However the Greek philosopher Plato's work, the Cratylus, is about language, and is largely centred upon items that are unmistakably words. There have been three main approaches to this problem. The first is to see the word as a semantic unit, a unit of meaning; the second sees it as a phonetic or phonological unit, one that is marked, if not by 'spaces' or pauses, at least by some features of the sounds of the language; the third attempts to establish the word by a variety of linguistic procedures that are associated with the idea that the word is in some way an isolable and indivisible unit.

Let us begin by the semantic definitions of the word. The word is said to be a linguistic unit that has a single meaning. The difficulty is of course in deciding what is meant by a single meaning, for meanings are no more easily identified than words. It is easy enough to show in a variety of ways that we cannot define words in terms of units of meaning. To begin with it is clear that very many single words cover not one but two or more bits of meanings. If *sing* has a single meaning, then presumably *singer* has more since it means *one who sings* and even *sang* must mean both *sing* and 'past time'. Conversely there are combinations of words in English that do not have separate meanings. *Put up with* cannot be divided into the three meaning units of *put, up* and *with*, but seems to have the single meaning of *tolerate*. The phonetic and phonological definition of the word states that it is possible to recognise a word by some feature of its pronunciation, as for example fixed stress. The problem is that for example *beat her* is often pronounced in exactly the same way as *beater*. It is possible to distinguish between *that stuff* and *that's tough* or *grey day* and *Grade A* but no distinction is seen between *a tack* and *attack*. The last idea, of considering a word as isolable and indivisible, comes from the American linguist Leonard Bloomfield (1933). The problem is what we are prepared to utter in isolation is almost certainly what we have learnt to recognise as a word in writing.

In conclusion, we have to say that the word is not a clearly definable linguistic unit. According to the descriptive level at which we place ourselves, phonetic-phonological, orthographic-graphemic, morphological, lexico-semantic or syntactic, a 'word' can take various characteristics, even and very often contradictory ones too. So, rather than trying to find the one and only one good definition of 'word', we will show rather how in different languages the concept 'word' appears in written form. To do this we could perhaps retake the concepts of 'words' and 'word forms' (the 'word' being the canonical representation of 'word forms' that are variants), but will this be sufficient? (Cardey et al. 2005).

In the Romance languages, in French:

> *pomme de terre*

is not recognised as three independent units but together as a single compound unit. Spanish on the contrary has compound units which resemble single units:

> *Saltamontes* (= grass hopper, locust): compound formed by concatenating the verb *salta* with the substantive *montes*

We find the following two types of composition in Romanian:

1. *untdelemn*: *unt* (butter) + *de* (of) + *lemn* (wood) which in English means oil
2. *floarea-soarelui* (= sunflower)

Encliticisation and contraction in the Romance languages are phenomena that have to be recognised.

The Slavic languages have the particular feature of attaching person marking morphemes (*–m, -ś, -śmy, -ście*) to different parts of speech:

> *Ciekawiśmy* = We are curious

The morpheme -*śmy*, first person plural attaches itself directly to the adjective, here the masculine plural *ciekawi*. The person marking morphemes can attach themselves in the same way to other parts of speech (personal pronouns, adverbs, interrogative pronouns, etc.).

In Turkish, compound words form a single word. When there is a vowel at the end of the first word and also at the beginning of the second, one of these disappears giving rise to one single form.

  Cuma ertesi   →   *Cumartesi*

In Arabic building a 'word' is based on a root composed essentially of two (atypical), three (triliteral, a very productive type), and four (quadriliteral) consonants within which vowels called thematic vowels are inserted. This root with its inserted vowels constitutes a 'scheme' or 'theme'. To illustrate Arabic morphological word composition, thematic vowels are going to be inserted into the root <*ktb*> which represents the global meaning 'write' to produce schemes (patterns) such as:

  kataba   (he wrote)
  kitaab   (a book)

Each root gives a kernel meaning:

  <*ktb*>   'idea of writing'
  <*rnn*>   'idea of ringing'

A particular feature of Arabic is the use of semantic doublets which are synonyms generally used at the end of sentences and which when translated often give a single word in the target language, as for example:

  la tu9add wa la tuhSa (neither evaluated nor counted)
      translation: 'which cannot be evaluated'
      and not: 'which can neither be counted nor evaluated'.

In some Asian languages, defining the concept of word is equivalent to wondering about that of the syllable. This latter is difficult to define because it varies according to the particular language's analysis.

In Thai a syllable requires three elements in order to form a semantic unit: an initial consonant, a vowel and a toneme. The final consonant is not obligatory.

In Chinese, two terms are linked to the concept of word: 字 (*zì*) and 詞 (*cí*). A 字 (*zì*) is a character (thus monosyllabic), whilst a 詞 (*cí*) can be monosyllabic, dissyllabic or polysyllabic. In other words, a 詞 (*cí*) can be composed of one or several 字 (*zì*). A 詞 (*cí*) is an independent semantic unit which can be translated by a 'word' in English and French whilst a 字 (*zì*) is sometimes a word and sometimes a morpheme (in which latter case it cannot function on its own). Furthermore, 詞 (*cí*) includes two notions: 單純詞 (*dān chún cí*: simple word) and 合成詞 (*hé chéng cí*: compound word). The majority of Chinese simple words are monosyllabic.

In Thai, simple words or fundamental words are predominantly monosyllabic. Each of them represents a single concept. A compound word is formed by the juxtaposition of single words (concepts) with no spaces between them:

/baj–1/ (leaf) + /klua:j–3/ (banana tree) = /baj–1+ klua:j–3/ (banana tree leaf)

The Thai vocabulary is formed principally by the juxtaposition of simple concepts. Two Thai nouns can be juxtaposed to compose a new noun. The 'head words' of Thai compounds are in general on the left and the dependent words extend to the right. However there exist a large number of terms borrowed from Pali and Sanskrit which use the reverse procedure. As in Chinese, if the order of words is reversed new meanings appear:

/pva:n–3//lu:k–3/: friend of (my) daughter or of (my) son
/lu:k–3//pva:n–3/: child of (my) friend (male) or of (my) friend (female)

Asian languages are written from left to right; nevertheless Japanese, Korean and Taiwanese Chinese can also be written from top to bottom. There are no separation marks except in Korean, and compounds are not separated. There are no capitals and the distinction between proper and common nouns is not simple.

All this shows that there are elements smaller than words which have some meaning.

One can say that in the knowledge of their language native French speakers, for example, find the 'word' *iront*, but one can also say that they construct it themselves by following a construction schema imposed on them by their language, here French, in the same way that the language imposes sentence construction. This leads to uncertainties as to what we should consider as significant units.

It is the development of comparative linguistics which has imposed the dissociation of the 'word' into more elementary significant units. Indeed the comparison of two different languages in order to establish their linguistic proximity cannot be done 'word' for 'word', but only by those elements to elements making up these 'words'.

What has also been decisive is the discovery of the proximity between the Indo-European languages and Sanskrit in which the internal plurality of the word is particularly striking, the different elements of the word being often juxtaposed in an evident manner (Panini 1966).

1.3.2   Morphemes and syllables

Morphemes are abstract (theoretical) units. They are represented phonetically and morphologically by morphs as the smallest meaningful but unclassified segments of meaning. If such morphs have the same meaning and a complementary distribution or if they stand in free variation, then they are said to be allomorphs of the same morpheme. For example in French, *all-* in *aller*, *ir-* in *iront*, *vai-* in *vais* and *aill-* in *ailles*, are the allomorphs of the verb *aller* (infinitive) for the latter's conjugation. In *doors, oxen* and *sheep*, *-s*, *-en* and ø are allomorphs of an abstract plural morpheme. Regarding the postulate of the unity of form and meaning, a distinction must be drawn between discontinuous morphemes, in which two or more morphs separated by other elements yield the morpheme's meaning as in German *ge* + *lieb* + *t*, where *ge-* and *-t* together mark the participle, and also so-called portmanteau morphemes (or contracted morphemes) in which the smallest meaningful segments carry several meanings as in *au* that is the contraction of the morphs *à* and *le* in French.

Concerning their semantic function we distinguish between lexical morphemes or lexemes, that denote objects, state of affairs, etc. of the extra-linguistic world and whose relations are studied in semantics and lexicology, and grammatical morphemes (inflexional morphemes) that express the grammatical relations in sentences and are studied in morphology and syntax. Regarding their occurrence or their independence, after Bloomfield we distinguish between free morphemes (roots or bases), which may have both a lexical function (*book, red, fast*) as well as a grammatical function (*out, and, it*), and bound morphemes in which it is a matter of either a lexical stem morpheme and inflexional morphemes (as in verb endings) or derivational morphemes of word formation (affix = suffix, prefix, infix).

Thus morphemes are not only grammatical words (prepositions, conjunctions, etc.) and affixes, but also tone, accent position and word order.

However the terminology is very variable from one language to another. American linguists generally use the word 'morpheme' with the same sense as the term moneme for Saussure's successors (3 monemes in re/tourn/ez) which should not be confused with morpheme (= grammatical moneme) in Martinet's terminology (2 morphemes or grammatical monemes re- and -ez).

1.3.3   Parts of speech

We now turn to the parts of speech. Trying to find some regular order within languages has resulted in tasks such as that of classifying what the Ancient Greek and Latin grammarians called the parts of speech (*tou logou, partes orationis*). The problem is that this leads to the concept of words being the smallest parts of speech.

Nevertheless, this division still exists to the present day. Of those who contributed to this division, Plato distinguishes the two classes of '*ónoma*' (= names; nouns) and '*rhêma*'(= statements; verbs), which represent noun and subject, and verb and predicate respectively, whilst Aristotle adds a third group, the 'indeclinables'. The Arabic grammarians founded their description of Arabic also by means of a classification of words in parts of speech, generally three: nouns, verbs and particles. Our current classification is based on the teachings of the grammarian Dionysius Thrax (first century BC) who proposes eight parts of speech: noun, verb, adjective, article, pronoun, preposition, adverb and conjunction. Such a classification was retaken by the Latin grammarian Aelius Donatus (IV century) in his *de octo orationis partibus* and has also subsequently been used by la Grammaire de Port-Royal (Arnauld & Lancelot 1660) (noun, pronoun, verb, participle, conjunction, adverb, preposition, interjection) and is still used with little change at the present day.

Is this classification valid for all languages? Even taking account of the comparison between Greek and Latin, two languages that are relatively close, one already sees divergences. Thus, Latin not having articles, the Latin grammarians when confronted with Greek introduced by force into their pronoun category the Greek article (*arthron*) and pronoun (*antonymia*) between which the Greek grammarians, such as Aristarchus of Samothrace, had carefully distinguished. If we consider other languages still further apart, it is difficult to envisage how some classification established from certain particular languages can be universal; the chances are slight that such a classification would be immediately adaptable to all other languages. Furthermore, the traditional classification in parts of speech has recourse simultaneously to different points of view. Marcus Terentius Varro (I century BC) in his *De lingua latina* uses morphological, combinatory and semantic criteria for his typology. The concept of the adjective was developed in the Middle Ages essentially to account for the fact that most adjectives designate qualities, and most substantives objects. Syntactic criteria being inadequate to distinguish them (in Latin, an adjective can be the subject of a verb) a compromise was devised by making them sub-classes of the noun category. It is clear that the distinction between *noun* and *verb* was based on the different roles played by these two classes in the enunciation activity. The former serves for designating objects and the latter for saying something about these objects, and this leads to the argument/predicate classification. However, this takes us away from the classification based on word classes because, for example, the *rhêma* function can be achieved by several other means than by using the verb in its grammatical sense, as the following examples of divergences in several languages show:

-   In Sanskrit, there are five parts of speech: noun, verb, adjective, adverb and preposition. There are neither articles nor relative pronouns, and this so too is the case in Chinese, Japanese and Korean.
-   In Turkish there are two major categories, verb and noun. Nouns can take many functions: adverb, interjection, pronoun, conjunction, etc.
-   In Chinese, Japanese and Korean, there are neither possessive adjectives nor relative pronouns.
-   In Thai, the relative pronoun is modelled on English grammar.
-   In Korean, to translate:
     la neige qui tombait (the snow was falling)
    the root of the verb is used to which is attached a relative ending. The relative pronoun is expressed by means of corresponding endings which are attached to the radical (root) of verbs of action, quality or state.
-   In Thai for expressing possession the preposition 'kh@:ng-1' can be used preceded by a noun and followed by a noun or pronoun.
-   In Asian languages the article does not exist. In certain cases, to indicate a term's semantic class, this absence is filled by means of classifiers (Cl), for example in Thai:
    -   /chan–5 + mi ;–1 + nang–5sv/
        me + have + book(s)
        Pron + V + N
        S + V + O
    -   /chan–5+mi ;–1+nang–5sv :–5+lem–3+nvng–2/
        me + have + book + 1 + Cl
        Pron + V + N + classifier for book + Num
        S + V + O + classifier for book + Num
-   Most of the Slavic languages have the same parts of speech as Latin. Compared with the Romance languages one notes the absence of both definite and indefinite articles.
-   Also in the Slavic languages, those units that do not correspond to the traditional grammatical categories are grouped together in a particle category:
    -   Interrogative particles (*czy* in Polish, *ли* in Russian)
    -   Expressive particles (*no, że* in Polish, *ну, же* in Russian) which are difficult to translate because they express attitudes and sentiments:
        No, no!   → Hey! Well, really!
        Иди же сюда!   → Come here!

Other studies have involved classifying based on semantic criteria; see the Grammaire de Port-Royal (Arnauld & Lancelot 1660). If the signifying element 'substance', common at the origin of the substantive nouns, has resulted in a substantive-signifier model enabling giving a (fictive) separate existence to what cannot even have such a separate existence (*la blancheur*, whiteness), it is because the substantive has as its primary origin the ability to appear in discourse in an autonomous way needing neither an adjective nor a verb. Compared with *blanc/*white, *blancheur*/whiteness does not call to mind, even in a confused way, the individual objects to which the word can be applied, and thus it can provide a sense by itself.

These few examples suffice to show the large divergences in parts of speech between different languages. The linguistic means used to provide certain functions are very varied and sometimes certain elements which are absent in one language are indispensible in another.

Syntactic analysis

Finally, we briefly review syntactic analysis.

We find in the Routledge dictionary of language (1966: 473) that syntax is:

> a system of rules which describe how all well-formed sentences of a language can be derived from basic elements (morphemes, words, parts of speech). Syntactic descriptions are based on specific methods of sentence analysis and category formation.

Here we will simply show the main syntactic analysis approaches from which many of the others are derived, and that these former, which at first view are apparently different, are in fact quite close to each other. In this respect, Vauquois (1975: 60–96) effectively shows how to pass from an immediate constituents binary tree to a dependency tree (we have modified some links). Given 3 nodes A, B and C in the immediate constituents element schema (see Figure 1) we consider A and B to be linked by an equivalence relation.

We can thus suppress node A and replace it with B obtaining the dependency element schema (see Figure 2) which is the dependency link where B governs C.

Application of the transitivity of this equivalence relation to the whole binary tree (see Figure 3) for the sentence:

Le but de cet exemple consiste surtout à illustrer divers graphes structuraux de l'analyse syntaxique

(The goal of this example consists above all to illustrate various structural graphs of syntax analysis)

leads very simply to the dependency structure (see Figure 4) of the same sentence.

Figure 1. Immediate constituents element

Figure 2. Dependency element schema

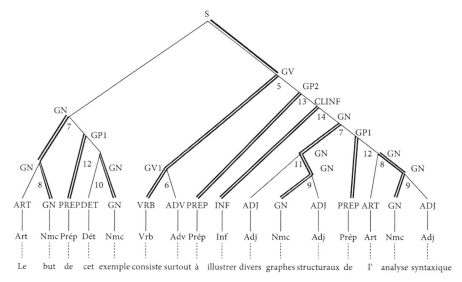

Figure 3. Immediate constituents binary tree

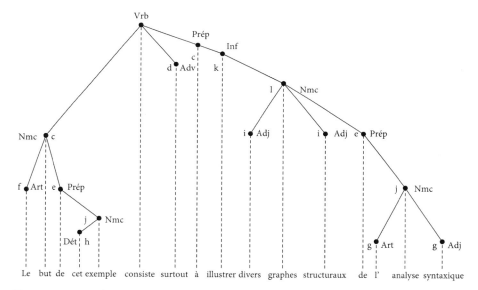

Figure 4. Dependency tree

Semantics

As has already been said, it is not possible to classify semantics at the same level as lexis, morphology and syntax. Concerning this point, let us see what certain linguists have written.

In his *Eléments de syntaxe structural* Lucien Tesnière takes care to distinguish between structural and semantic plans. We provide in what follows a translation of what he wrote (Tesnière 1982: 40-42):

> The structural plan is that used for formulating a thought's linguistic expression; it depends on grammar to which it is intrinsic. On the contrary, the semantic plan is specific to a given thought, an abstraction of every way to express the thought linguistically. The semantic plan does not depend on grammar to which it is extrinsic, but only on psychology and logic. [...] The structural and semantic plans are thus theoretically entirely independent of each other. The best evidence of this is that a sentence can be semantically absurd at the same time as being perfectly correct structurally. [...] In practice the two plans operate in fact in parallel; the structural plan serves no other object than rendering the thought's expression possible, this by means of the semantic plan.

From the speaker's point of view, it is the sense which he wants to give to his message which causes him to use such or such structure; he thus follows an onomasiological path, from the sense to the forms which express this sense. As for the hearer, he follows a semasiological path, from forms to the sense or from the message as a datum to an interpretation, this being decoding the message.

We provide in what follows a translation of what Jean Dubois (1965) writes concerning "*ses principes de la méthode distributionnelle*".

> For describing the rules of a code, one cannot take the sense as a basis. Distributional analysis disregards semantics as an access to the code's structures as a means for investigation and analysis. Distributional analysis does not reject the importance of sense in the message's construction.
>
> For Hjelmslev a language system is a purely formal reality; it is the collection of abstract relations between its elements, independently of any phonetic or semantic characterisation of these [...] Thus for Hjelmslev and Z. Harris, when one analyses a language, the sense is put between brackets and it is used to verify the identity or non identity of the utterances, and nothing more. [...] Such a scientific attitude supposes that the messages, utterances and code rules exhibit a certain degree of stability.

CHAPTER 1.6

Norm in language

1.6.1 Synchrony and diachrony

Passing from some concept to reality is not simple when one speaks of norm. First of all we have to situate ourselves in time because languages are likely to change.

No system exists outside time. "When boundaries retreat indefinitely or are vague, one ends up with the aorist duration of proverbs or that of theoretical propositions" (translated from Gentilhomme 1985: 51).

Where does the investigative field of grammar start and end? It is here where the two terms synchrony and diachrony occur.

> *Synchronic linguistics* will be concerned with logical and psychological connexions between coexisting items constituting a system, as perceived by the same collective consciousness.
> *Diachronic linguistics* on the other hand will be concerned with connexions between sequences of items not perceived by the same collective consciousness, which replace one another without themselves constituting a system. (Saussure 1922: 140, English translation: 98)

> The two words 'synchronic' and 'diachronic' include the base 'chron' borrowed from Ancient Greek which evokes a period of time. Synchrony embraces systems as they function here and now on a horizontal plane so as to ensure perfect understanding between those who speak. Synchronic grammar thus has for objects language states. Diachrony has as objective describing, by way of going back through the course of time, the accidents which have contributed to altering the systems. The description of the state of a language and the study of the accidents, these two contributions, far from being opposed to each other, are complementary. (Translated from Robert Léon Wagner (1973: 14)).

The chronological factor is abstracted from synchronic linguistics which latter involves studying a language taken as a whole from within some community which speaks it here and now. In contrast, diachronic linguistics involves instances in time for studying the evolution of a language. A linguistic phenomenon is said to be synchronic when all the elements and factors that concern it belong to one and the same moment in time and one and the same language. A linguistic phenomenon is said to be diachronic when all the elements and factors that concern it belong to different states of development of one and the same language.

One has often emphasised the exclusively historical character which has marked linguistics for the whole of the XIX century and the beginning of the XX century. [...] . What has been new, from the Saussurian point of view, has been to understand that language itself contains no historic dimension, that language is synchrony and structure. (Translated from (Benveniste 1966: 5)).

Saussure shows that a language, at every moment in its existence, ought to appear as an organisation, as a system, and that what justifies its synchronic aspect is its integration in the whole, in the system of language. The synchronic connexions can be established without taking account of diachronic considerations, but it can turn out that the former are in conflict with diachonic connexions. Firstly certain synchronic connexions are not diachronically justified. In synchrony, for example, in French one has the connexion '*léguer-legs*', a connexion which is analogous to '*donner-don*', '*jeter-jet*', etc. However, there is no historical connexion between *léguer* and *legs* which latter is connected to *laisser*.

Vice versa, many established historical connexions have no synchronic reality and this is because they cannot be incorporated in present day language. We find, for example, in French the adverbs in 'ment' formed from 'grièv', 'nuitante' and 'scientre', all these latter no longer exist.

The conclusion commonly held amongst linguists is that linguistic evolution can have systems as starting and ending points, and as such the evolution is described by means of the transformation of one synchronic structure to another synchronic structure.

Historical linguistics explains a word q by a word p preceding q in time if passing in time from p to q is the particular case of a general rule valid for other words, and thus makes it to be understood that p' became q', p" became q", etc. This regularity implies that the difference between p and q is due to such or such of their constituents and that in all other words where this constituent appears, it will be changed in the same way.

[...]

In linguistic domains other than phonology, attempts at establishing a 'history of systems' have seen little progress with the exception of semantic field analysis developed by J. Tiers on some part of the German lexis. (Translated from Ducrot & Todorov (1968: 183-187)).

However, Ferdinand de Saussure speaks of "the autonomy and interdependence of synchrony and diachrony" (Saussure 1922: 124, English translation: 87).

Grevisse (1980: 28) writes of descriptive grammar and historical grammar (here translated into English) "Descriptive grammar explains the linguistic usage of a group of humans at some given epoch. It is ordinarily limited to noting and recording the 'good usage', that is to say the constant usage by those who take care to speak well and write well."

Historical grammar studies language development and transformation. It explains the changes which are produced in the linguistic usage between two epochs more or less distant from each other. Historical grammar juxtaposes the descriptive grammars of several successive epochs and looks for the relations existing between the grammatical features that are observed there; for example, in German the different phonetic changes which have transformed *Gasti* to *Gäste* and *Handi* to *Hände* but have not reached the grammatical feature itself, the duality of the singular and the plural.

Thus before doing anything else, studying a language involves collecting together a set of utterences as varied as possible emitted by the users of the language at a given point in time. One then tries to find regularities in the corpus so as to produce a description which is both ordered and systematic, and avoiding that this be just a simple inventory.

However, Saussure says of this: "Language at any time involves an established system and an evolution. At any given time, it is an institution in the present and a product of the past. At first sight, it looks very easy to distinguish between the system and its history, between what it is and what it was. In reality, this connexion between the two is so close that it is hard to separate them." (Saussure 1922: 24, English translation: 9).

1.6.2 Good usage

> Vulgar spoken language continues its assured underground journey; it flows like living water under the rigid ice of written and conventional language; then one good day the ice breaks, the tumultuous flow of popular language invades the still surface and brings anew life and movement. (Translated from (Bally 1952: 13)).

The term 'descriptive' has within its definition a reference to 'normative'. The norm being 'good usage', thus the problem of what is properly said and what is not is posed.

During the XVII century in France as in numerous countries, etiquette governed over both politics and morals. Vaugelas and his disciples imposed etiquette on the French language; grammar became a catalogue of faults and an amusing entertainment for Society. Grammar is a register of usages, or rather, good usages, the quality of the usage being above all judged according to the quality of the user.

Who dictates good usage? "For some conscious innovation to penetrate syntax, morphology or pronunciation, there is need of the complicity of a class or elite." (translated from Bally (1952: 118)).

> Good usage is that of important writers, journalists and authors of manuals of the epoch. These are the models of the cultivated public whose actions counterbalance the unconscious construction of language's history by the anonymous speaking mass. From Vaugelas to Grevisse, these are the guarantors that the arbiters of language in France invoke. (Translated from Hagège (1985: 191)).

Saussure (1922: 125–127, English translation 87–89) and Hjelmslev compare language to a game. Hjelmslev writes:

> We can say that a game structure is a collection of rules indicating the number of pieces in the game and the way each of the pieces can be combined with the others, and this is different from language usage. To describe the usage of the game, it would be necessary to give instructions, not so much concerning the way one plays, but in fact the way one has played up to now. (Translated from Hjelmslev (1968: 207)).

H. Frei (1929: 32) sought less to describe language than the way language functions, that is to say the way language is used at a given epoch. For this reason his study concerns not only language said to be 'correct' but also all what is out of place in relation to traditional language such as faults, innovations, popular language, slang, unusual cases, contentious cases, grammatical perplexities etc. These disparities reveal what the speaker is expecting from the language in not finding what is wanted. The principal linguistic needs would be:

- assimilation which leads to uniformising both the sign system (analogical creation) and also elements that follow each other in the discourse (grammatical agreement);
- brevity (ellipsis, compound words);
- invariability;
- expressivity (the speaker seeks to mark his discourse with his personality, from which a perpetual invention of figures of speech).

All these functions explain the faults but also numerous aspects of 'good usage' (constituted by the faults committed in the past).

In discourse we often find completely new structures or terms mixed with others declared outmoded, but kept in usage by a certain social inertia. This is the reason why we talk about fuzzy zones.

> Wherever and whenever it imposes itself, good usage ought to conciliate two contradictory tendencies, that is to say, it has to maintain during three or four generations a state of equilibrium between resistances of a conservative character and innovative dominances. Within a synchronous state, these inverse forces oppose each other, and they act there in the person of individuals of different ages and backgrounds who participate in this state. Their incessant conflict in natural idioms disorganises the system little by little, itself always temporary and fragile. (Translated from Wagner (1973: 11)).

PART 2

Modelling the norms

In this part of the book, starting with the linguistic model, we develop the mathematical model, based on the concept of the norm, which is constructive and which is based on model theory. This model serves not only the linguist but also as a means of communication between the linguist and those involved in applications and associated procedures (linguists themselves, software engineers, quality and other engineers, etc.), in providing capacities for capturing linguistic performance/competence, exhaustivity, traceability, verification, validation, case based testing and benchmarking, rigorous qualification procedures and so on.

Model

Faced with all these difficulties linked to variously typology, domains constituting language analysis, the concept of word, parts of speech, problems of delimitation, the only model possible is a model which can be adapted according to specific needs.

We now briefly review what can be a model.

Models, whatever be their type, are expressed by means of formal languages which are more or less abstract; we provide in what follows a translation of Durand's distinction into three major categories of language (Durand 1987: 55):

- *Literary languages* formed of literary symbols assembled in structures. The most widespread of these languages are evidently the everyday languages: English, French, German, etc., but it is also necessary to mention technical languages which are almost universally widespread as well as computing languages which are continually being created and developed.
- Iconic or pictographic languages formed of graphic symbols distinguished according to several criteria including size, colour, intensity and orientation, and forming varied structures in the form of variously networks, tables or curves. Graphs and matrices are the most used forms for this type of language, at least in the Occident. There are other forms of iconic language; these are pictograms of which the most well known are Chinese ideograms and Egyptian hieroglyphs [...]
- *Logico-mathematical languages* formed of abstract symbols obeying precise structuring rules; the most important of these depending on set theory which appears as the basis of mathematics.

Contrary to the mathematical concept of set, a system for the grammarian and the linguist contains simultaneously constituents as well as the elements and the relations susceptible to be established between all the considered constituents and elements.

Our model

This chapter explains the foundations of systemic and micro–systemic linguistic theory that we have developed, together with its algebraic and computational aspects. This theory is itself based on model theory, on the concept of system and above all on the concept of micro–system. We explain the linguistic model as well as the corresponding mathematical model, how each of these models functions, and above all how the two models are associated to form a single model allowing not only fine-grained and detailed analyses of language(s) but also their generation together with the production of algorithms for eventual applications.

2.2.1 The linguistic model

2.2.1.1 Macrocosmic representation

It seems to be impossible to present a language as a system in its entirety because such a system would not be humanly prehensible.

Because objects are sometimes too complex to be described in a rational manner in one single step, it is easier to proceed by stages using a progressively finer scale.

Language being too 'dim' and furthermore too complex to be analysed in its entirety, just as is the solar system which latter is impossible to delimit precisely and which varies too with time, we endeavour to represent language as a system taking into account its fuzzy aspects. To this end Figure 5 shows how our system can be situated in time.

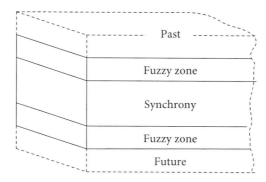

Figure 5. Our system situated in time

We provide a translation of what Rober Léon Wagner writes (Wagner 1973: 55): "Whatever the epoch, literature provides evidence concerning the state of a language both in the process of its disappearing as well as its becoming established".

What is not so easy is to model the way the parts composing a language are interrelated. Our macrocosmic or macroscopic approach to this problem functions as shown in Figure 6.

In Figure 6 we see that though semantics appears, it is not part of the macrocosm's internal functionality, but that senses are either analysed or generated by the macro-system.

Our representation is in line with Benveniste (1966: 12) who wrote (here translated into English) "It is thus possible to envisage several different types of both description and formalisation of language, but all of these must be based on the assumption that language contains meanings, and because of this, it is structured".

One can ask why represent the macrocosm, that is the 'galaxy' of morphology, syntax and lexis, our macro-system, in this way where there are parts which overlap? Rather than answering by means of a lengthy argumented justification, we will instead give some examples drawn from French which will be more explicit (see Figure 7: French lexico-morpho-syntactic system).

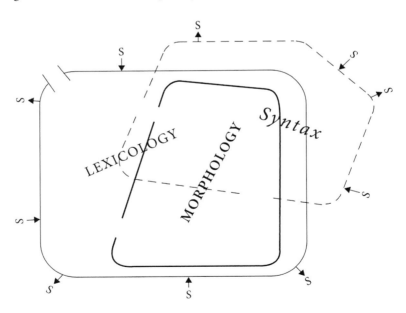

S = semantics

Figure 6. Macrocosmic approach to modelling the way the parts composing a language are interrelated

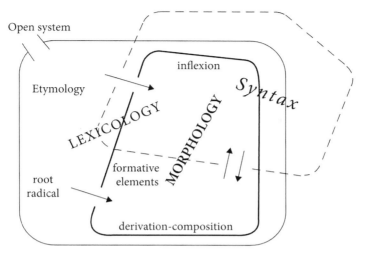

Figure 7. French lexico-morpho-syntactic system

We start seeing that the 'planetary' elements of our 'galaxy' include smaller elements which move according to their needs from one 'planet' to another.

We see that the roots and radicals can travel towards the inflexions, the formative elements, etc.

To be more concrete, we now provide some chosen examples of units of each of these elements that our planets are composed of and see how for some given senses the units travel to one planet and/or another in order to collect other elements.

If we want the French radical '*beau*' to take the form of an adverb, '*beau*' has to enter the inflexion element to change gender (from masculine to feminine) to become '*belle*', and then '*beau*' transformed into '*belle*' has to enter the derivation-composition element to take the suffix '*ment*'; see Figure 8.

Now let us see how the radical '*fleur*' becomes '*fleuristes*', a term which is the same category as '*fleur*'. '*fleur*' must find the derivational element '*iste*' and then '*fleuriste*' must find the inflexion marker for the plural so arriving at '*fleuristes*'; see Figure 9. (However the plural can be marked in different ways, this according to the radical. In consequence our representation would need to be refined further).

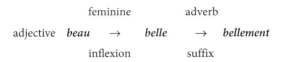

Figure 8. French radical *beau* to take the form of an adverb

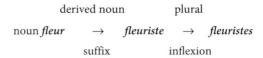

Figure 9. French radical '*fleur*' becomes '*fleuristes*'

Finally, we will now see how the radical '*achet(er)*' is transformed when it is in the environment '*les pommes qu'elle a achetées*'. '*achet*' has to pass through two elements, that of inflexion and that of syntax; see Figure 10. We start to see that there is a problem as inflexion and syntax are not at the same level. Thus it would be necessary to look closer at how our system functions with its various elements.

Figure 11 recapitulates how our system functions for the examples with radicals *beau, fleur* and *achet(er)*.

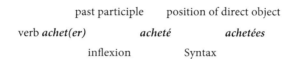

Figure 10. French radical 'achet(er)' transformed in the environment '*les pommes qu'elle a achetées*'

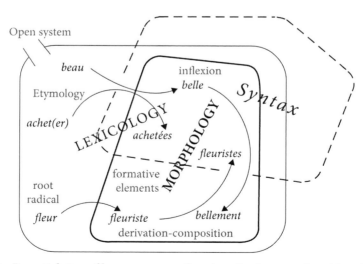

Figure 11. Recapitulation of how our system functions for the examples with radicals *beau, fleur* and *achet(er)*

2.2.1.2   Microscopic approach to morphology

With the help of our 'microscope', we are now able to show how the sub-elements of our system function. Though we have discussed lexis, syntax and morphology, we are now going to see that in reality such a separation is not viable.

We are going to show with examples how one can represent the morphological system which, as we have already said, does not exist independently of lexis and syntax.

As the terminology used for language analysis is so very variable, before advancing further, we need to establish the terminology that we will be using, but firstly let us see what linguists write concerning this matter.

The majority of comparativists distinguish two component types inside the 'word' (it is clear that what are being studied here are inflexional languages). The first component type contains components which are called in German 'Bedeutungslaute' and in the French grammatical tradition 'sémantèmes' (semantemes) or 'radicaux' (radicals) or according to Hjelmslev 'pleremes'. The second component type contains components called 'Beziehungslaute' and morphèmes (morphemes) (or grammatical markers designating variously thought categories and the intellectual points of view imposed by the mind or by reality). Concerning morphemes it has become usual to distinguish between inflexions (which belong to the conjugation and declension systems) and affixes (which are more independent one from another). Furthermore, according as to whether an affix appears before or after the semanteme, it is considered to be a prefix or a suffix, and certain languages also have infixes within the semanteme.

Whilst retaining the idea of the necessity for decomposing the 'word', most linguists nowadays do not accept the above classification alleging that it is at best suitable for the languages of classical antiquity, and that it has been introduced for modern Indo-European languages by the projection of the past into the present, this being contrary to purely synchronic descriptions.

American linguists designate these components with one and only one term, 'morpheme', and European linguists speak variously of 'morphemes' and 'formants'.

A.J. Greimas (1986) divides the morphological classes as follows. Certain of the classes are open, the number of word-items constituting them not being limited (nouns, verbs); inflexional analysis is part of the morphological categories inventory, and that of roots of the lexical description inventory. Concerning roots, Gustave Guillaume (1973: 117) writes (here translated into English) "The radical takes position in the word as an integral element [...]. The root however takes position in the word not because of being followed (or preceded) by the formative elements, but with the purpose and ability to place these latter within itself". Other classes are on the contrary closed (pronouns, determiners, adverbs, etc.); the number of commutable items for each of these classes is indeed very limited. All their words are generally considered as being grammatical rather than lexical.

Martinet (1973) distinguishes:

- grammatical monemes (such as 'present indicative' or 'indefinite article') which belong to closed inventories in the sense that the appearance of a new tense or new article will lead necessarily to modifications to the existing tenses or articles' values.
- lexical monemes 'which belong to open inventories' where the appearance of a new foodstuff noun for example will not lead necessarily to a change in the value of 'soup'.

In the Grand Larousse de la langue française (1977, article «grammaire», p.2284) we find (here translated into English):

> It seems that at any instant one can add a noun to the set of nouns and even a suffix to the set of suffixes without affecting the values and the uses of other nouns or suffixes. However, adding a new 'number' to the couple singular/plural, a new article in the system in which there is the opposition *le* and *un* (*the* and *a*) will on the contrary cause a significant reorganisation of the whole. Thus conceived, the criterion of classes that are either open or closed enables including affixes within the lexis. However it is far from providing the expected clear-cut frontier.

André Martinet (1973: 119) writes (translated into English) "Lexical monemes are those that belong to inventories without limits. Grammatical monemes are those which, in the examined positions, alternate with a relatively reduced number of other monemes".

However this distribution is not decisively pertinent seeing as it leaves unanswered the question about affixes which, though not great in number, do, even so, constitute a relatively open class.

For Martinet, 'lexeme' includes both simple base units and also affixes, which he finally classes amongst the lexemes, even given the relatively closed set of the affixes' constituent elements, and this leaves open the perspective concerning the constitution of new affixes. B. Pottier (1962) distinguishes between on the one hand the purely lexical lexemes, the base roots or simple terms, and on the other hand the formants which are the affixes. The compound and derived units are situated at an immediately superior level in the morpho-syntactic hierarchy, and these make up the lexical syntagms.

Emile Benveniste (1966: 119) writes (translated into English) "The list of morphological categories, though seemingly so varied, is not unlimited. One could imagine a sort of logical classification of these categories which would show their arrangement and their transformation laws".

The variations in form have given rise to what are called systems such as 'number' and 'tense' (inflexional systems). Gustave Guillaume (1973: 223) writes (translated into English):

> Language is a system of systems, a systemised assemblage of nested container systems (having their own content of interior positions), one nested in another and which differ from each other in many points of view except for their notion of container; each of these systems is included in a larger one, the largest of these being the language itself.

2.2.1.2.1   *Nested elements*

As we have already said that it is not possible to represent a language in its entirety, we continue our process of decomposition with our microscope and examine in greater detail one of our planets, for example morphology, knowing that we cannot detach it together with how it functions from the other planets. This will lead us to speak of morpho-syntax and lexical morphology.

Figure 12 shows the representation that we propose for our French morphological system.

As linguists do not all agree amongst each other concerning terminology, we give the sense of each linguistic term that we use in associating the term with a linguist or school, as we have explained in the preceding paragraphs.

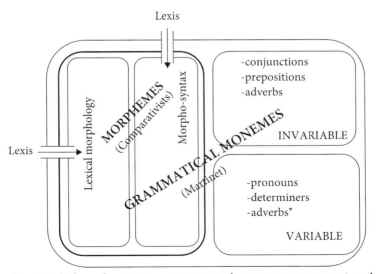

*E.g. *tout* before a feminine noun starting with a consonant or an aspirated *h*

Figure 12. French morphological system

Recalling certain linguistic terms, grammatical monemes after Martinet are the inflexions together with those which alternate in considered positions with a relatively reduced number of other monemes (closed classes) such as adverbs and determiners. Martinet classes the affixes in the lexis and it is for this reason that we keep the comparativists' sense of morpheme for inflexions and affixes with which we are more in agreement.

What is important is rather to know how these elements are grouped together rather than the name that we give them.

We observe that our system is composed of four parts of which three communicate with each other.

If we now superimpose our two schemata Figure 7: French lexico-morpho-syntactic system and Figure 12: French morphological system, we observe the two major components of morphology: morpho-syntax and lexical morphology (Figure 13).

We now see the reason for the overlapping of our three systems, morphology, syntax and lexis.

We continue our investigation by going inside one of our two major components of morphology, that of morpho–syntax (the other being lexical morphology).

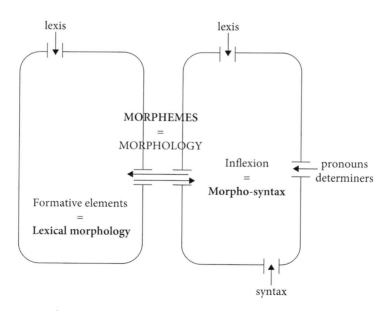

Figure 13. Morpho-syntax and lexical morphology

2.2.1.2.1.1 *Morpho–syntax.* In this section we give examples drawn from French which we will subsequently examine in further detail.

We have already used the term 'inflexion'. In French this term is used for expressing variously gender, number, case, person, tense, mood and voice. A schema of the French morpho-syntactic system is shown in Figure 14.

A lexical classification based on inflexion appears relevant as not all 'words' can accept the same inflexions or inflexional endings.

Gender applies to nouns, adjectives, determiners, pronouns and verbs; case uniquely to certain pronouns; person to certain determiners, to verbs and to pronouns; tense, mood and voice to verbs. Number applies to the five groups (determiners, pronouns, nouns, adjectives, verbs).

We thus develop our schema, which is shown in Figure 15 and represented in tabular form in Figure 16.

Again we see why the systems that we started with, lexis, syntax and morphology are interrelated and thus overlap with each other.

We now need to continue our exploration of each of the above mentioned groups but before doing so we have to distinguish between two types of inflexion: 'internal' inflexions (also called morpho(pho)nemes by N.S. Troubetzkoy (1976)) which modify the base itself, and 'external' inflexions (or radical inflexions) which join the endings directly to the base. Figure 17 shows the schema of the French morpho-syntactic system further developed with the elements of the components or sets.

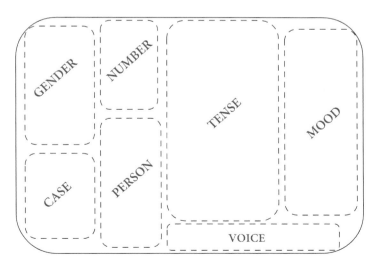

Figure 14. Schema of the French morpho-syntactic system

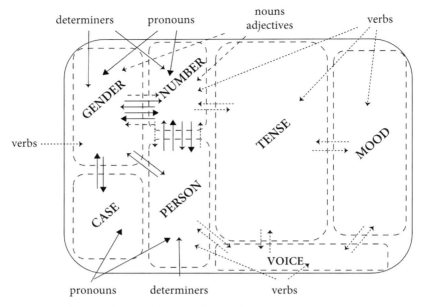

Figure 15. Developed schema of the French morpho-syntactic system

Inflexion / Word category	gender	number	case	person	tense	voice	mood
adjectives	X	X					
determiners	X	X		X			
nouns	X	X					
pronouns	X	X	X	X			
verbs	X	X		X	X	X	X

Figure 16. The French morpho-syntactic system in tabular form

Certain linguists and grammarians are not in agreement concerning the role of the French conditional, which for some is not a mood but rather a 'hypothetical future'.

As in our schema shown in Figure 7: French lexico-morpho-syntactic system we will see, but this time more precisely, how and why the lexical elements travel from one set to another recuperating the elements that are necessary for their insertion within the language (see Figure 18).

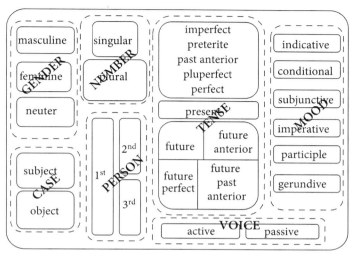

Figure 17. Schema of the French morpho-syntactic system showing the elements of the components or sets

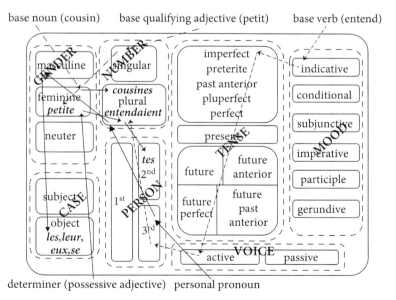

Figure 18. French morpho-syntactic system showing element recuperation for insertion within the language

In Figure 18 we see for example that for the word *cousin* to become *cousines* it has to recuperate the feminine mark followed by the plural mark. Unfortunately things are not so simple. For instance the morpheme for plural is not always produced in the same way, and this is also so within the same lexical category (for example the adjectives).

Now, if we want to enter for example the 'plural' set, it is necessary to assemble firstly all the morphemes used to mark the plural in French, and secondly classify the bases according to the category of which they are members (nouns, verbs, qualifying adjectives, pronouns and determiners). As an example, the French qualifying adjective plural morpheme set is shown in Figure 19.

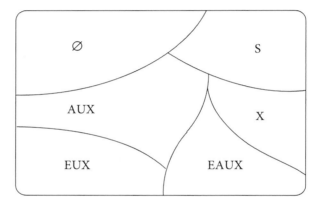

Figure 19. French qualifying adjective plural morpheme set

We then need to know what the conditions are for using such or such morpheme (or ending). We know for example that those adjectives ending in 'al' generally form their plural with 'aux' but that there are exceptions.

To advance further in our exploration of how language functions in respect of morpho–syntax, we will need to use the concept of micro–system.

We are now going to explore the second major component of morphology, that of lexical morphology.

2.2.1.2.1.2   *Lexical morphology*
We have already seen that the French lexical morphology component comprises two other components: derivation and composition (see Figure 7).

2.2.1.2.1.2.1   *Derivation.* Derivation is termed explicit derivation and implicit derivation. Implicit derivation can give words new functions without there being any change to their external appearance. In respect of explicit derivation, a derived word is a lexical unit which is formed from a base by the addition of a special ending.

Three categories of endings can be distinguished which serve for forming:

- nouns and adjectives, that is nominal derivation
- verbs, that is verbal derivation
- adverbs, that is adverbial derivation

Derivation can also involve the removal of a word's final syllable, that is, back formation.

2.2.1.2.1.2.2 *Composition.* Composition serves for forming lexical units (LU) (we prefer this term to 'words') either in combining simple lexical units with already existing LUs, or by preceding simple LUs with syllables which themselves have no proper existence. (Certain linguists think that it is more normal to consider derivations to be all those units that are formed by adding an affix (suffix or prefix) to a base; thus 'impossible' is seen to be a derivation and not a compound (Mounin 1974)).

Composition exists too involving sentence members and even complete propositions.

The French lexical morphology component can be represented as shown in Figure 20 and in more detail in Figure 21.

We can also have compound derivations by parasynthetical formation as in en+col+ure (neck) in French.

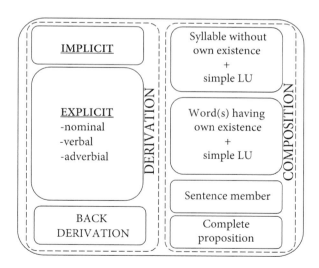

Figure 20. French lexical morphology system

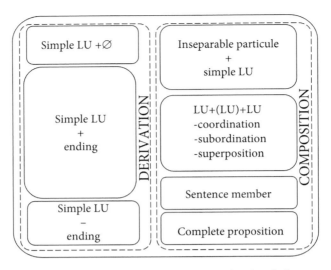

Figure 21. French lexical morphology system in greater detail with formation modes

In Figure 22 we show examples of elements passing from one sub–component to another.

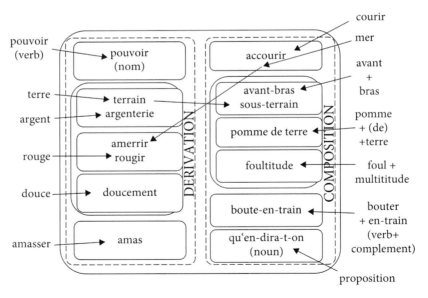

Figure 22. French lexical morphology system with examples of elements passing from one sub–component to another

The endings as well as the different particles and lexical units in respectively derivation and composition are numerous. For this reason, as for the morpho–syntactic
component, we will have to use the concept of micro–system to be able to progress
further into the functioning of lexical morphology.

2.2.1.3   Systemic linguistic modelling of other languages

In this section we show in brief systemic linguistic modelling examples for Malay
and Arabic.

2.2.1.3.1   *Malay.* Using our systemic linguistics model, we now see how a component of an agglutinative language, such as Malay, functions (Mohd Nor Azan Bin
Abdullah 2012). The result of the analysis implies a relation between the system of
affixations and the verbal system in Malay; this is shown in Figure 23.

The system for Malay works as follows. The base of the word enters the open
system. If we want a verb formed from an adjective, we have to circulate it in the
formative elements system in case it ought to take a prefix. For example, if we take
the adjective '*sakit*', we want this adjective to take the prefix '*me*' in order to become a verb. The adjective '*sakit*' combines with the prefix '*me*'. According to the
rules of formation for this prefix, the initial letter '*s*' of the word '*sakit*' must be
changed into '*ny*' for nasal harmonisation. So the prefix '*me*' becomes '*meny*' and

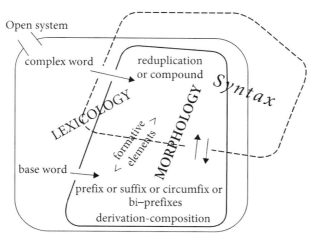

Figure 23. Model of the systemic linguistic analysis that implies the relation between the
system of affixations and the verbal system in Malay

agglutinates with the rest of the letters of '*sakit*'. Then we combine the result with the suffix '*i*' in the suffix system. The final result is the derivation of the adjective '*sakit*' into the verb '*menyakiti*' (*to become painful*). To know how the system functions, we need to understand how word formation (derivation) works.

The same process of word formation can be applied to explain the formation of compound verbs. We take for example '*berperikemanusiaan*'. This verb is made up of two nouns: '*peri*', the first one is combined with another noun '*manusia*'. According to Figure 23, the two nouns enter the compound word formation system and so they form '*perimanusia*' but this happens to be insufficient. Then this new noun enters the affixation formative system as it needs two prefixes (*ber-* for '*peri*' and *ke-* for '*manusia*') and it also needs a suffix (an) at the end of '*manusia*'. These all combine together to become an intransitive verb: '*berperikemanusiaan*'.

We give another example with a bi-prefix: *mem+per.* (*memper-*). '*memperkenalkan*' is first obtained from the verb '*kenal*' (to know), it then enters the affixation formative system and it takes a bi-prefix '*memper-*' with '*mem*' giving an active sense to the verb and '*per-*' permitting the verb to be transitive. So, the bi-prefix permits the verb to be active and transitive at the same time; it becomes '*memperkenal*' but this happens to be insufficient. It therefore needs a suffix '*–kan*' to become '*memperkenalkan*' (to introduce).

2.2.1.3.2   *Arabic.* We show by means of an example due to Mohand Beddar the micro–systemic representation of agglutination in Arabic (Figures 24 and 25), together with the equivalence in English (Figure 26).

So as to advance in our decomposition and especially see the detailed functioning of the new sub-components, we have recourse to the concepts of micro–system and of algorithm.

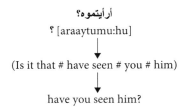

Figure 24. Example for the micro–systemic representation of agglutination in Arabic

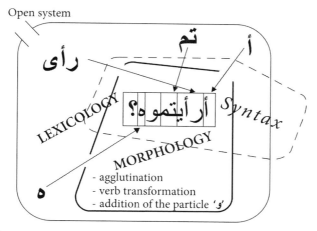

Figure 25. Micro–systemic representation of agglutination in Arabic- example: ؟أرأيتموه: ؟[araaytumu:hu]

أ	Interrogative particle	Is it that
رأيْ	Verb رأى (*see*) 2nd person plural perfective	have seen
تم	Personal pronoun suffix 2nd person plural (subjet)	you
و	Linking particle	
ه	Personal pronoun suffix 3rd person singular (direct object)	him
؟	Question mark	?

Figure 26. Equivalence Arabic–English for the example: ؟ [araaytumu:hu]

2.2.1.4   Concept of micro–system

A micro-system is a finalised system, small enough, even predegenerating, to be treated sufficiently approximatively, in a realistic time delay, according to discursive or experimental methods, available, but large enough to show its purpose. The size of a micro-system depends on the final objective given at the outset.

2.2.1.4.1   *Algorithmic micro–system.* Algorithms

> "His augrim-stones layen faire a-part
> On shelves couched at his bedded heed"
> (Chaucer, The Milleres Tale)

Augrim = Middle English from old French augori(s)me from Medieval Latin algorismus, itself the deformation of the name of the mathematician (Muhammad ibn Mūsā) al-Khārezmi, who published a book circa 820 on arithmetic. The word was modified due to the influence of the Greek word 'arithmos', that is, 'number'.

"By an *algorithm* is meant a list of instructions specifying a sequence of operations which will give the answer to any problem of a given type." (Trakhtenbrot 1960: 3).

> Broadly speaking, an algorithm is a procedure which performs a particular process and which must terminate, and which, for some particular source (input) data enables obtaining target (output) data in the form of a developed result depending functionally on the source data. The outcome of the procedure is the transformation of objects belonging to a source set which is relatively large or indeed infinite to objects belonging to a target set. We note that frequently an algorithm can be considered as the explicitisation of some operator acting on a variable of some particular domain. (Translated from Gentilhomme (1985)).

The role played by algorithms in mathematics requires that they necessarily possess the three following properties:

1. Determinism. Randomness is excluded in the choice of operations which must be clear and unequivocal. Identical initial data lead to identical results, whoever the user is.
2. Generality. This property concerns the fact that any object belonging to some determined class can be an initial datum for some problem that is solved by means of an algorithm.
3. Finality. The algorithm is always aimed on obtaining a sought for result which, given suitable initial data, will always be obtained. It is possible that for certain initial data, the algorithm is not applicable in which case the execution of the algorithm either terminates without giving a correct result or never terminates.

Concerning Finality, this leads to the fundamental question of what can be calculated algorithmically, and thus mechanically, and what cannot? There are indeed problems that are undecidable. No algorithm, which by definition must terminate, can be formulated to solve them (Turing 1936). See for example (Aho & Ullman 1995: 772–773 Undecidability) and (Trakhtenbrot 1960: 86–88 Algorithmically unsolvable problems).

This notwithstanding, "determining what types of problems can be solved or not by algorithmic means is of epistemological interest because it enables estimating the degree of generality of such and such problems, and such and such ways of thinking." (translated from Landa (1965)).

The most useful algorithm is the one which is the most rational and which takes the least time.

Landa mentions the following characteristics:

- produce a result (efficiency)
- finite number of instructions (finiteness)
- clear instructions (intelligibility)
- elementary instructions (elementarity)
- operator independent result (objectivity)
- unique solution (unicity)
- same result if one starts again (fidelity)
- set–wise solution and not for particular cases (generality)

that can be regrouped under the three properties defined above: Determinism, Generality and Finality.

Descartes wrote that it is necessary to split up, disintegrate each difficulty up to extracting its '*molécules*', then its '*atomes*'. "*...diviser chacune des difficultés que j'examine en autant de parcelles qu'il se pourrait et qu'il serait requis pour les mieux résoudre*" (Descartes 1637), in English "divide each of the difficulties that I examine in as many parts that it could and would be best for solving them". For certain levels, these "*atomes ne sont plus seulement des faits, mais, aussi, des relations entre les faits*"–these atoms are not only facts but also relations between facts.

2.2.1.4.2 *Examples of micro–systems.* We are going to show how to apply the concept of algorithmic micro–system by means of four examples. We will see examples of more complex applications such as machine translation, sense mining and still others in the third part of the book.

At this point in the exposition we are at the stage of the morpho-syntactic set or component (see Figures 18 and 19) and the lexical-morphology component (see Figure 22). We need data in order to pursue the analysis and also some objective seeing that our system ought to be adaptable according to the goal to be reached. The goal to reach could be, for example, to provide the agreement of the French past participle, or to know when to double the consonants in French for words starting with 'ap' or indeed in English before the endings -ed, -ing, -er, -est, -en, or, for yet another language, the problem of declension in German.

We now turn to our first algorithmic micro–system example.

2.2.1.4.2.1 *Algorithmic micro-system example 1 and its various representations.* French words starting with 'ap'.

We start with the most simple of the problem examples: when to double the consonant in French for words starting with 'ap'.

Our theory requires that everything be treated in intension. The theory thus respects the property of Generality which means here that any object which is the member of some determined class itself represented by an object (we say canonical) can be the initial datum for some problem.

Seeing that our theory requires working in intension rather than in extension, methodologically the first thing to do is to build a representative corpus of all the cases which can appear in the language.

In the set of examples in the corpus, the accidental deviations or divergences will constitute the variants to the norm. It is thus necessary that the norm and the accidental divergences or variants appear and that sometimes these latter be described anew in terms of micro–systems. We will see that sometimes we need sets defined not by means of rules (comprehension, that is in intension) but rather constituted by lists (enumerations) of lexical units ('words'), that is, in extension. Thus generalities and particularities will form our micro–system which in turn can be decomposed into smaller micro–systems or call on other micro–systems.

It is here that the concept of norm also plays its role. Grammarians and linguists have observed languages for a long time, and have formulated grammatical rules which explain how languages work at least for those wishing to learn them, or for children learning their own language.

We shall thus start from such grammars (we have, in the case of French, consulted no less than 26 grammar manuals; they are indeed very numerous and variable in terms of their completeness), and from a corpus containing all the cases expressed by the grammars, and have thus built our source data.

In the numerous pages of rules that we have read and that we will not reproduce here, we have found, it is true, much information, but not one of the grammar manuals has really provided what we wanted, be this due to a lack in terms of ease of use, or a lack in terms of exhaustivity. Sometimes we managed to extract data in the form of rules, but in reality, from the outset the data tended to be somewhat chaotic.

2.2.1.4.2.1.1 *The algorithm.* Having constructed our system using data from the grammar manuals together with the data that we ourselves have had to add or reformulate, we can now present our first example 'When to double the consonant in French for words starting with 'ap''.

We give in Figure 27 a representative corpus which is sufficient as it contains all the possible cases, that is the canonical elements representing all the different classes of the partition of the set of all the French words beginning with 'ap'.

For reasons that we have already explained, we will not use diachrony, that is, evolution of language features in time; our system works in synchrony.

We now give the reformulated grammatical rule from which the algorithm will be constructed (Figure 28).

Now we have to define the optimal procedure, this in terms of the nature and order of instructions, knowing that we will have a system of questions for which the responses are either 'yes' or 'no'.

a-ache	a-eler	a-latir	a-ostropher
a-aisement	a-endre	a-laudir	a-othéose
a-aiser	a-entis	a-licateur	a-ôtre
a-anager	a-ercevoir	a-lomb	a-péritif
a-araître	a-ériteur	a-ocalypse	a-réhender
a-areil	a-erture	a-ocope	a-robation
a-areillé	a-etisser	a-ocryphe	a-rocher
a-arence	a-étit	a-ogée	a-rofondir
a-arté	a-euré	a-oplexie	a-roximation
a-artement	a-icale	a-orter	a-ui
a-artenir	a-icole	a-oser	a-urer
a-athie	a-iculteur	a-ostasier	
a-eau	a-itoyer	a-oster	
a-el	a-lanir	a-ostiller	

Figure 27. Corpus for 'When to double the consonant in French for words starting with 'ap''

Verbs as well as the 'words' of their family starting with 'ap' take 2

'p' except for:

apostasier, apostropher, apostiller, aposter, apercevoir,

apanager, apetisser, apeurer, apurer, aplanir, apitoyer, aplatir,

apaiser

as well as the 'words' of their families (or their derivations)

'appartement' and 'appétit' take 2 'p'

Figure 28. Reformulated grammatical rule for 'When to double the consonant in French for words starting with 'ap''

For this we have to classify all or part of the instructions in a hierarchical or ordered manner, so that when these are put into operation, the problem will be solved in an automated way. Everyone and anyone ought to be able to find the solution. Everything is determined in advance; the user becomes an automaton but someone who has to be able to answer 'yes' or 'no' to questions. Thus it is necessary that each instruction be clear, simple and elementary so as to be understood by anyone. There is no room for chance; two different users ought to arrive at exactly the same result.

As we have already mentioned, the only authorised user answers are 'yes' or 'no', that is, true or false.

The procedure as a process always has a start. At this point, it is necessary to verify that the material, here the French 'words' starting with 'ap(p)', conform to the expected procedure, here the processing of French 'words' starting with 'ap(p)'.

One can start thus:

- S(start): verify that one has a 'word' starting by 'ap(p))'
- Q represents the conditional questions or instructions intended to ensure the presence of a criterion and to switch within the procedure in consequence. Each question is uniquely labelled.
- P represents the operator to apply to the considered object. This is also uniquely labelled.
- F indicates that the procedure has finished

It must not be forgotten that it is necessary to process the particular before the general. For example here the list of exceptions, for which the general rule is not applicable, ought to be processed before the verbs.

We now write our algorithmic procedure for French 'words' starting with 'ap(p)' which is shown in Figure 29.

0S0 = ensure that one has a word starting by "ap(p))"

1Q1 = Is the word a verb or one of its derivations (word of its family)? If yes go to Q2, if no go to Q5

2Q2 = Is the verb or one of its derivations (word of its family) in the list of exceptions? If yes go to P3, if no go to P4

3P3 = put 1 "p". F. (Finish).

4P4 = put 2 "p". F. (Finish).

5Q5 = Is the word "appartement" or "appétit"? If yes go to P4, if no go to P3

Figure 29. Algorithmic procedure for French 'words' starting with 'ap(p)'

This manner of processing avoids having to refer to all what is not a verb, that is, the complement of verbs and words of their families with respect to the French lexis, which by default is all the rest. This allows us to process by default a word taking one 'p' like "apache" which is neither a verb nor derived from a verb, nor "appartement" nor "appétit", and which thus takes 1 'p'.

We now see how we can represent our algorithm using different forms of flow diagram.

2.2.1.4.2.1.2 *Representations*. Reading an algorithmic procedure usually requires close attention; one can easily lose oneself in the intricacies of the instructions. Schemata can aid understanding. A flow diagram is a schema which enables visualising a procedure, and in particular an algorithmic procedure.

One can devise schemata which are more or less convenient according to what one has to do; one should not confuse algorithms with their flow diagram visualisations.

We now see how to adapt the notation devised by two Russian mathematicians, A.A Ljapunov and G.A. Sestopal, which has the advantage of being linear and thus economical in terms of space. This notation can be used for algorithms as large as one wishes.

The principles concerning the notation are as follows:

– conditions (conditional instructions 'Q' as described above) are indicated, that is, labelled by lower case Latin letters (a, b, c...), and the operators (processing instructions 'P' as described above) by upper case Latin letters (A, B, C...);
– a full stop '.' indicates that the algorithm has finished its work (F as described above);
– an indexed rising arrow (e.g. $\uparrow^1$) indicates to go to the descending arrow with the same index (e.g. $\downarrow^1$).

The sequence of symbols is read as follows:

– from left to right, and stop reading at the first full stop '.';
– when an upper case letter (operator) is read, execute the corresponding order (process);
– when a lower case letter (condition) is read, if the reply (for the corresponding condition) is 'yes', pass to the following letter; if the response is 'no', follow the rising indexed arrow just after the lower case letter, find the descending arrow with the same index, and read the letter that follows.

The data for our algorithm for French 'words' starting with 'ap(p)' is shown in Figure 30.

Conditions (conditional instructions 'Q')
    a.   verb or derivative (word of the same family)
    b.   list of exceptions
    c.   appartement, appétit

Operators (processes 'P')
    A. p
    B. pp

Figure 30. Data for the algorithm for French 'words' starting with 'ap(p)'

We show the algorithm in Ljapunov and Sestopal's flow diagram representation (adapted by us) in Figure 31.

$$a\uparrow^1\, b\uparrow^2 B.\downarrow^2 A.\ \downarrow^1\, c\ \uparrow^3 A.\downarrow^3 B.$$

Figure 31. Algorithm for French 'words' starting with 'ap(p)' in Ljapunov and Sestopal's flow diagram representation

One reads if 'a' is true read 'b', if 'b' is true, perform the operation indicated by 'B'. The full stop '.' indicates that the algorithm has finished its work. If 'a' is false follow the rising arrow with index 1, read the letter after the descending arrow with the same index, here 'c', if 'c' is true read the letter that follows, here 'A' and apply the operation indicated by 'A'.

We can also represent our algorithm by means of a binary tree where 'a' indicates that condition 'a' is TRUE, and 'ā' FALSE (see Figure 32).

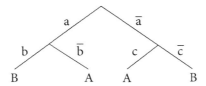

Figure 32. Algorithm for French 'words' starting with 'ap(p)' in a binary tree representation

The same algorithm can thus be paraphrased in many different ways and transcoded to varied semiotic systems without in any manner losing its algorithmic virtues.

We now show another representation which is due to Landa (1965, 1976) and adapted by us.

Principles:

- find and label the pertinent criteria that are evoked: a, b, c;
- find and label the processes: A, B;
- construct a table with all the possible sequences (combinations) of the criteria;
- amongst these sequences, find those which really correspond to the problem to be solved and associate them with the operators implied by them.

From this point it is straightforward to obtain all the functionally equivalent algorithms.

In Figure 33 we show Landa's representation that we apply for French 'words' starting with 'ap(p)'. In Figure 33:

- The value 1 means that the corresponding condition is satisfied, that is, has the logical value TRUE.
- The value 0 means that:

either

the negation of the corresponding condition is satisfied, that is, has the logical value FALSE

or

that this condition has not been envisaged at this level

Line number	Values Criteria (Conditions)			Implies ⇒	Process (Operator)
	a	b	c		
1	1	1	1		
2	1	1	0	⇒	B
3	1	0	1		
4	1	0	0	⇒	A
5	0	1	1		
6	0	1	0		
7	0	0	1	⇒	A
8	0	0	0	⇒	B

Figure 33. French 'words' starting with 'ap(p)'–Landa's representation

The number of cases can be very large depending on the number of conditions. The procedure can become unwieldy and be difficult to apply without computational approaches.

For the problem of French 'words' starting with 'ap(p)', for the $2^3 = 8$ possible cases, only 4 are kept by the algorithm (Figures 33), the 5 others ought not to appear in reality.

We now turn to our second example.

2.2.1.4.2.2 *Algorithmic micro-system example 2.* Doubling or not of the final consonant in English words ending with -ed, -ing, -er, -est, -en.

We follow the same approach as before.

After studying grammar manuals we have as data:

- 'words' of a syllable of the form CVC (consonant, vowel, consonant) double the final consonant
- 'words' terminated by CVC with the last syllable accented double the final consonant
- 'words' terminated by CVC or by CV(pronounced)V(pronounced)C terminated by -l or -m double the final consonant in England except for "(un) parallel", otherwise the previous rule applies
- "handicap", "humbug" double the final consonant everywhere
- "worship", "kidnap" double the final consonant in England
- 'words' terminated by -ic take -ck
- "wool" doubles the final consonant in England, but not elsewhere

An algorithm can be devised and represented in what we call our 'organigram' representation due to the author (Cardey 1987) and which is shown in Figure 34.

a. 'word' of a syllable of the form CVC   D (doubles the final consonant)

b. 'words' terminated by CVC or by CV(pronounced)V(pronounced)C   N (does not double the final consonant)

c. last syllable accented   D

d. 'word' terminated by -l or -m   N

e. in England   D

f. "(un)parallel"   N

g. "handicap", "humbug"   D

h. "worship", "kidnap   N

e. in England   D

i. 'word' terminated by -ic   ck (c becomes ck)

j. "wool"   N

e. in England   D

Figure 34. Organigram representation of the algorithm for Doubling or not of the final consonant in English

The organigram representation is read thus. Suppose that we wish to write "model(l)ing" in British English; ought there to be one "l" or two "ll"? We have:

a is not true, so go to b;
b is true, so go to c (c is an exception of b);
c is not true, so go to d;
d is true, so go to e;
e is true if we write in British English, so go to f;
f is not true;
so we apply the operator corresponding to the last true condition which here is e and we therefore apply D;
therefore we write "modelling" with two "l" ("ll").

This sequence, which we can write:

$\bar{a}$ b $\bar{c}$ d e $\bar{f}$ which is: no a yes b no c yes d yes e no f

represents a proof and is also a trace of the execution.

We can represent the same algorithm by means of a binary tree. Figure 35 illustrates such a tree with the execution trace emphasised in **bold**.

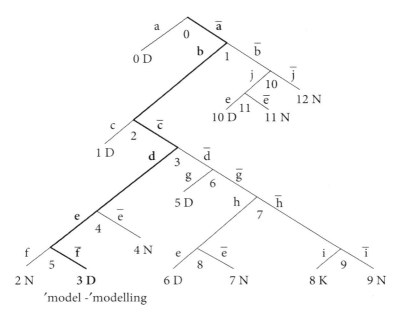

Figure 35. Binary tree representation of the algorithm for Doubling or not of the final consonant in English with trace

We now look at a more complex problem, our third example.

2.2.1.4.2.3 *Algorithmic micro-system example 3.* The agreement of the French past participle.

In French there are 54 different cases for the agreement of the past participle.

In Figure 36 (read the first column, then the second), we see the apparently complete algorithm's organisation. We say 'apparently' because this algorithm, which we call the 1st algorithm, is in reality linked to another algorithm, the 2nd (Figure 37), which determines how to do the agreement according to the form of the lexical items to which it is linked and which latter can be simple or compound amongst others. The 1st algorithm (Figure 36) is shown without operators; we show in detail an example of its use further on in our exposition.

We take as an example:

une foule d'hommes que j'ai **vu?** sortir
(a crowd of men that I have seen going out)

In our example, **vu** is a past participle in canonical form, and 'sortir' is an infinitive.

We now take our microscope to see how to solve this problem at a finer level.

We can extract from the preceding 1st algorithm (Figure 36) the micro–algorithm which will process French past participles followed by an infinitive.

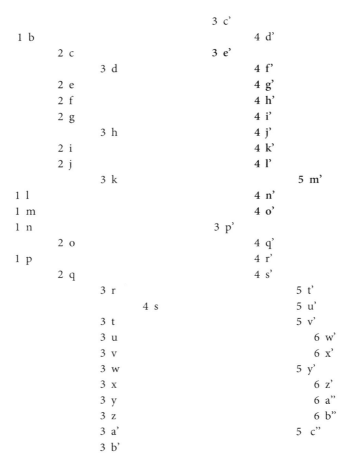

Figure 36. The agreement of the French past participle: 1st algorithm (read in the order first column, then the second)

The micro–algorithm starts at condition e' of the preceding algorithm up to condition o' (marked in **bold** in Figure 36) for the first part of the solution, and continues with the 2nd algorithm (Figure 37) to obtain the complete solution.

$$a \uparrow^1 b \uparrow^2 G. \downarrow^2 c \uparrow^3 M. \downarrow^3 G. \text{ or } M. \downarrow^1 d \uparrow {}^4 e \uparrow^5 G. \text{ or } M. \downarrow {}^5 M. \downarrow^4 f \uparrow^6 g \uparrow^7 F.$$
$$\downarrow^7 E. \downarrow^6 h \uparrow^8 H. \downarrow^8 i \uparrow^9 j \uparrow^{10} I. \downarrow^{10} I. \text{ or } J. \downarrow {}^9 k \uparrow^{11} L. \text{ or } M. \downarrow {}^{11} C.$$

Figure 37. The agreement of the French past participle: 2nd algorithm which enables the agreement according to the lexical items on which it depends

After simplification, the micro–algorithm for the first part of the solution is as shown in Figure 38.

Algorithm

a.   the past participle is preceded by "**en**" → **I**
b.   the past participle of the verb **faire** is immediately followed by an infinitive → **I**
c.   the past participle of the verb **laisser** is immediately followed by an infinitive → **I** or **A**
d.   **eu, donné, laissé** are followed by "**à**" and the direct object **refers** to the infinitive → **I**
e.   the verb (past participle) marks an opinion or a declaration (**List**) → **I**

f.   the direct object performs the **action** of the **infinitive** → **A**
g.   the direct object **refers** to the **past participle** → **A**
h.   the direct object **refers** to the **infinitive** → **I**

List: affirmé, assuré, cru, dit, espéré, estimé, nié, pensé, prétendu, promis, reconnu, voulu

Operators

I :   invariable
A :   agreement with the direct object

Figure 38. Micro–algorithm for French past participles followed by an infinitive–first part of the solution

To obtain the result, one passes by not a, not b, not c, not d, not e, f is TRUE, one thus applies A = agreement with the direct object.

However one has now to continue to find the form of the direct object, by means of the 2nd algorithm which gives at condition k as operator 'could agree with "foule" 'feminine' agreement or "hommes" 'masculine and plural agreement' and thus **vu?** is 'vue' or 'vus'.

Before turning to the mathematical model, we look at our fourth and final example which is drawn from another language, German.

2.2.1.4.2.4   *Algorithmic micro-system example 4.* Declension of German adjectives
The problem to be solved with its solution is without doubt one of the most difficult to formulate and represent and this is due to its complexity.

This problem will allow us to introduce the concepts of 'canonical form' (out of context) and 'form in context', and above all the system of relations that acts between these two concepts.

We are going to see the representation of the algorithmic system which describes the relationship between the form of a German adjective in context and its canonical form (Cardey & Greenfield 2000).

We have three sets that we are going to call systems due to the way they function:

- System 1: the German adjectives in their canonical form;
- System 2: their inflexions;
- System 3: the relationship that structures the German adjective system 1 according to the inflexional system 2.

We now give an extract of a corpus which is representative of all the cases of the declension of German adjectives (see Figure 39).

> etwas warme Speise
>
> sie möchte mehr rote Rosen
>
> in welcher aufregenden Stunde
>
> welcher andere Text
>
> welche reizende frau
>
> beider jungen Menschen
>
> fern von aller spöttischen Überlegenheit
>
> bei allem bösen Gewissen
>
> die Spitzen einiger grossen Radnägel
>
> die Filme einiger guter Filmemacher
>
> die Filme einiger guten Filmemacher
>
> einiger poetischer Geist
>
> nach…einiger erfolgreicher Zurwehrsetzung
>
> einiges milde Nachsehen
>
> einiges geborgenes Mobiliar
>
> by einigem guten Willen
>
> einigen poetischen Geistes
>
> einiges poetischen Geistes (rare)
>
> etliche schöne Bücher
>
> die Taten etlicher guter Menschen
>
> die Taten etlicher guten Menschen
>
> etliche schöne getriebene Becher

Figure 39. Declining German adjectives: extract of a corpus representative of all the cases of the declension of German adjectives

As for the previous examples, the data have been organised in an algorithmic manner.

An extract of System 3 in which System 1 is embedded is shown in Figure 40.

Source System 1 is arranged as a result of target System 2 and the relationship System 3. Target System 2 (operators) is also arranged as a result of relationship System 3 and is shown in Figure 41.

c(Condition_Name,Condition).

:

c(1110,   'the adjective is preceded by: the following words without an
article: einzeln, gewiss, zahlreich, verschieden, übrige, derartig,
letztere, obig, selbig, sonstig, etwaig, ähnlich, besagt,
sogenannt, gedacht, ungezählt, unzählbar, unzählig, zahlos,
zahlreich, weiter(e)').

c(1170,   'the case is: the dative masculine or neuter or the genitive
plural').

c(1231,   'the adjective is preceded by: a cardinal or an ordinal').

c(1232,   'the adjective is preceded by: etwas or mehr').

c(1233,   'the adjective is preceded by: a personal pronoun').

c(1234,   'the adjective is preceded by: a demonstrative or relative
pronoun').

c(1235,   'the adjective is preceded by: an invariable word').

c(1236,   'the adjective is preceded by: an indefinite adjective').

c(1237,   'the adjective is preceded by: an article or by a determiner
declined on the model of the article').

c(1238,   'no determiner precedes the adjective').

c(1300,   'the cardinal or the ordinal is preceded by: an article or a
determiner declined on the model of the article').

c(1450,   'the case is: the dative singular masculine feminine or neuter,
or the nominative plural').

c(1550,   'the case is the dative singular').

c(1700,   'the indefinite adjective is declined').

c(1741,   'all').

c(1742,   'ander').

c(1743,   'einig').

c(1744,   'etlich or etwelch').

c(1745,   'folgend').

c(1746,   'irgendwelch').

c(1747,   'manch').

c(1748,   'mehrere').

c(1749,   'sämtlich').

c(1750,   'solch').

c(1751,   'viel').

c(1752,   'wenig').

c(1753,   'an indefinite adjective other than: all, ander, einig, etlich or
etwelch, folgend, irgendwelch, manch, mehrere, sämtlich,
solch, viel, wenig').

:

Figure 40. German adjective declension system: conditions (extracted from System 3–part of)

o(Operator_Name,Operation).

o(180,	'the adjective does not change but it can decline').
o(500,	'the -c disappears when declined').
o(800,	'the adjective rests as such').
o(850,	'the -e preceding the inflexion disappears').
o(950,	'the -e preceding the inflexion can be removed').
o(1090,	'the -e preceding the inflexion or the –e of the inflexion can be removed
o(1490,	'add -en or add the inflexion of the definite article').
o(1850,	'add -en or -e to the adjective').
o(1940,	'the inflexion -en will be used more frequently than -em').
o(2040,	'add -er to the adjective').
o(2100,	'add -e or -es to the adjective').
o(2160,	'add -en to the adjective').
o(2230,	'add -en or -er to the adjective').
o(3120,	'add -e').
o(3145,	'the adjective takes the inflexion of the definite article').

Figure 41. German adjective declension system: operators (System 2)

We also have operators which are themselves conditions which invoke other different operators as for example for the different declension types, strong and weak. The data associated with these declensions are in their turn represented algorithmically with conditions and operators. We show this by means of a simple example where condition c(1235) invokes algorithmic operator a(3310) whose associated algorithm gives as final operator o(2160), this taking place in what we call Super–system 4, an algorithmic part of this being shown in Figure 42.

In the initial algorithm, there are 121 possible paths which lead to the different inflexions.

A given path can process a set of adjectives whose size (cardinality) can go from one single element (for example '*hoch*' which is kept 'apart' because it loses the 'c' when declined) to a great number of elements as for example the default rule.

A part of the Super-system 4 showing its content is given in Figure 43.

If for example we have the adjective **einig** (its canonical form) and we want to arrive at **einigen** (an inflected form), we have source (condition) 1236, we find 1700 then 1743 then 1960 and 2120 with target 2160 which is **einig** + **en**. What is interesting is that having the source system, the target system as well as the system of relations and our super-system, not only can we go from the source to the target, but we can also go from the target to the source, as shown in Figure 44.

```
root_algorithm(a(10)).

a(10,
:
        c(1231)          -> a(3310)
              c(1300) -> o(2160)
                    c(3210) -> o(3120)
        c(1232) -> o(3145)
              c(3310) -> o(2160)
        c(1233) -> a(3310)
              c(1450) -> o(1490)
        c(1235)-> a(3310)
        c(1236) -> a(3310)
              c(1700) -> a(3210)
                    c(1741) -> a(3210)
                          c(1810) -> o(1850)
                    c(1742) -> a(3310)
                          c(1900) -> o(1940)
:
).

a(3310,
        c(3310) -> o(2160)
        c(3311) -> o(3145)
).
```

Figure 42. German adjective declension system or Super–system 4: algorithm (part of)

For example, suppose that we start from the target system and want to find the source items leading to the target. For the target o(3120), Figure 44 shows part of the system of links which leads to this target. Starting from operator o(3120), we can go up to a(10) by passing by c(3080),c(3040),c(3000),c(1237), or as well by passing by c(3210),a(3210) (algorithm), c(1753),c(1700),c(1236). It is also possible to extract or find a missing link, an element, and bring this to the fore, as is for example shown in Figure 45.

If one does not know which determiner to use in a text, the system will be able to find it, this due to the descriptions produced by the system. Thus our analysis, which started with the problem of the German adjective declension system, has enabled the construction of another system, that for the German determiners.

In our example, the data missing from the initial German adjective declension system is **einig, etlich, etwelch, folgend, mehrere** and **viel**.

:

c(1236,'the adjective is preceded by: an indefinite adjective') ->
            a(3310)
        c(1700,'the adjective is declined')-> a(3210)
            c(1741, 'all') -> a(3210)
                c(1810,'the case is: the nominative plural')
                        -> o(1850)
            c(1742,'ander') -> a(3310)
                c(1900,'the case is: the dative singular
                    masculine or neuter') -> o(1940)
            c(1743,'einig') -> a(3310)
                c(1960,'einig- is used in the singular')->
                        a(3210)
                    c(2000,'the case is: the nominative
                        masculine or the genitive
                        or dative feminine') -> o(2040)
                    c(2060,'the case is: the nominative
                        or accusative neuter') ->
                            o(2100)
                    c(2120,'the case is: the dative
                        masculine or neuter') ->
                            o(2160,'add -en to the
                                adjective')
                c(2190),'the case is: the genitive plural')
                        -> o(2230)

Figure 43. German adjective declension system: Super-system 4 with content (part of)

:

[o(2160),c(3310),c(1110),a(1110),c(700),a(10)]
[o(2160),c(3310),c(1110),a(1110),c(1110),a(10)]
[o(2160),c(3310),c(1232),a(10)]
[o(2230),c(2600),c(1743),c(1700),c(1236),a(10)]
[o(2230),c(2600),c(1744),c(1700),c(1236),a(10)]
[o(2230),c(2600),c(1745),c(1700),c(1236),a(10)]
[o(2230),c(2600),c(1748),c(1700),c(1236),a(10)]
[o(2230),c(2600),c(1751),c(1700),c(1236),a(10)]
[o(*3120*),c(3080),c(3040),c(3000),c(1237),a(10)]
[o(*3120*),c(3210),a(3210),c(1753),c(1700),c(1236),a(10)]
:

Figure 44. German adjective declension system: target to source (part of)

One can envisage applications of this approach in spelling and grammar correctors and for other applications too such as 'Studygram' (Cardey & Greenfield 1992), our system which explains how to solve problems as we have seen with the

$$[o(2230),c(2600),c(1743),c(1700),c(1236),a(10)]$$
$$[o(2230),c(2600),c(1744),c(1700),c(1236),a(10)]$$
$$[o(2230),c(2600),c(1745),c(1700),c(1236),a(10)]$$
$$[o(2230),c(2600),c(1748),c(1700),c(1236),a(10)]$$
$$[o(2230),c(2600),c(1751),c(1700),c(1236),a(10)]$$

leading to:

$$[o(2230),c(2600),c(17..),c(1700),c(1236),a(10)]$$

|

'einig'
'etlich' or 'etwelch'
'folgend'
'mehrere'
'viel'

Figure 45. German adjective declension system: extract or find a missing link, an element, and bring this to the fore

agreement of the French past participle or the doubling or not of the final consonant in English problem. We will see in Part 3 Methodologies and Applications that the Studygram methodology has also been used for creating dictionaries (the MultiCoDICT system (Cardey, Chan & Greenfield 2006)).

2.2.1.5 Our model for syntax

Up to now we have been concerned with morpho–syntax; we now turn to how to represent syntax which can also concern sets of lexical units (compounds), or sequences often called syntactic sequences, and also morphology as we have already seen.

Amongst the micro–systems that we have already discussed and which are constructed according to particular needs, certain of these will be used according particular languages, so we will need a micro–system for the grammatical phenomenon of 'case', for example in Arabic, German and Russian.

We know now too that these micro–systems will themselves have need of sets organised variously as morpho–lexical, morpho–syntactic and lexico–morpho–syntaxic systems.

We are now in the organisation of our galaxy with some stars attracting others but also some which repel others, and furthermore some stars that come alive whilst others are dying.

We can thus represent a structure which given a sentence or a part of a sentence will be able to provide a sense.

Language and languages also present structural norms, so how can our model give prominence to these? The first thing to do is to find these norms; we remark that there has been very little work done in this area.

2.2.1.5.1   *Syntax and morphology.* We take again our example of the French past participle followed by an infinitive; however this time we want a representation which will be able to generate the agreement automatically and even correct an existing erroneous agreement.

Given the problem and the language, we take the micro–systems that we need for our example; see Figure 46 (we will see other such examples in Part 3 Methodologies and Applications).

In Figure 46 we see that the elements needed to solve the problem are varied and that they do not belong to one and the same of the traditional divisions but rather to several of these, or to subsets of these which are organised ready to be used in aggregates which we call 'structures'.

Thus we ought to be able to find those particular aggregates or structures where such a past participle could be found. We have six possibilities a', b', c', d', k' and j' (see Figure 47). In what follows, the elements appearing between parentheses are optional. This means that in a' (see the following), which represents the maximal possible structure:

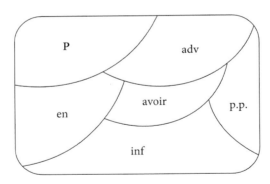

P	represents all that can be a noun or a nominal group of one or more elements
adv	means adverb
p.p.	means past participle
prep	means preposition
inf	means infinitive
avoir	is the conjugated auxiliary of 'avoir'

Figure 46. French past participle followed by an infinitive: elements needed to solve the problem

   a'. Pl(,)(adv)**en**(P3)(adv)(P4)(adv)avoir(adj)(P3)(P4)(adv)p.p.(adv)
      (prep)(P5)inf(P5)

one finds

   z'. Pl **en** avoir p.p.inf

which represents the minimal aggregate, as well as other sentences containing certain elements of the maximal structure.

All sentences whose structure contains at a minimum the elements of z' and at
a maximum the elements of a' in the order indicated in a' will have their past participles processed in the same way, that is 'I' ('invariable'), that is there is no agreement of the past participle.

All other aggregates are exterior to this system.

Knowing also that each and every aggregate follows a precise order, any aggregate containing the same elements but which does not respect the order will be
eliminated.

Using this model, we can now construct all the sequences which will enable
formulating our algorithm which in turn ought to respect a processing order;
see Figure 47. We observe that the algorithm's order must be strictly respected
except for c' and d' which can be inverted with no change in the result.

     Algorithm:
      a'. Pl (,)(adv)**en**(P3)(adv)(P4)(adv)avoir(adj)(P3)(P4)(adv)p.p.(adv)
          (prep)(P5)inf(P5) → I
      b'. P1(,)(adv)(P2)(adv)(P3)(adv)**en**(adv)avoir(adj)(P3)(adv)p.p.
          (adv)(prep)(P5)inf(P5) → I
      c'. Pl(,)(adv)P2(adv)(I')(P3)(adv)(P4)(adv)avoir(adj)(P3)(P4)(adv)**fait**
          (adv)(P5)inf(P5) → I
      d'. Pl(,)(adv)P2(adv)(I')(P3)(adv)(P4)(adv)avoir(adj)(P3)(P4)(adv)**List**
          (adv)(prep)(P5)inf(P5) → I
      k'. Pl(,)(adv)P2(adv)(I')(P3)(adv)(P4)(adv)avoir(adj)(p3)(P4)(adv)
          p.p.(adv)(prep)P5(adv)inf → I
        u'. P5 is the direct object → **agreement with P2**
          g'. P2 = "que" → **agreement with Pl**
          h'. if P3 exists → **agreement with Pl**
      j'. Pl(,)(adv)P2(adv)(I')(P3)(adv)(P4)(adv)avoir(adj)(P3)(P4)(adv)p.p.
          (adv)(prep)inf(adv)P5→ I
        u'. P5 is the direct object→ **agreement with P2**
          g'. P2 = "que" → **agreement with Pl**
          h'. if P3 exists → **agreement with Pl**

     List: affirmé, assuré, cru, dit, espéré, estimé, nié, pensé, prétend, promis,
     reconnu, voulu

     Operator **I: invariable**

Figure 47. Agreement of the French past participle: automatic algorithm

The execution of such algorithms is performed in the following manner:

> If **a'** is true then the past participle is **invariable** (operator **I**);
> If **a', b', c', d'** are false and **k'** true then if **u'** is true then if **g'** is true then make the agreement with **Pl**; otherwise, **g'** being false then if **h'** is false apply the operator of the last true condition, **u'**, that is to say, make the agreement with **P2**.

With the algorithm we can, for example, obtain the correct response (agreement or no agreement) for a sentence such as:

> Ces bûcherons, je   les   ai   fait? sortir
>   P1,   P2 P3   avoir p.p.   inf
> (These lumberjacks, I have made them leave)

If we examine the algorithm, we find that the structure shown at condition **c'** is the appropriate one:

> **c'.** Pl(,)(adv)P2(adv)(I')(P3)(adv)(P4)(adv)avoir(adj)(P3)(P4)(adv)**fait**(adv)
>   (P5)inf(P5) → **I**

which provides the correct response, operator **I**, this being that there is no agreement of the past participle.

Figure 48 gives examples of sentences which could enter structures within the algorithm.

> Les mesures qu'il a voulu prendre
> Ces arbres, je les ai vu abattre
> Les comédiens qu'on a autorisés à jouer
> On l'a vue courant dans les bois
> Des hommes que l'on avait envoyés combattre
> Des hommes que l'on avait envoyé chercher
> Ces bûcherons, je les ai vus abattre des chênes
> Les peines que nos parents ont eues à nous élever
> Que j'en ai vu sortir sous les huées !
> Je les ai fait chercher partout
> Je les ai fait jouer
> Mes chers collègues, je vous ai laissé parler
> Une chambre qu'elle leur avit dit être le petit salon
> Les volcans que j'ai eu à nommer

Figure 48. Agreement of the French past participle: examples of sentences which could enter structures within the algorithm

Take for example:

Ces bûcherons, je les ai   vu? sortir des chênes
    P1,   *P2 P3 avoir p.p. inf*   *P5*
(These lumberjacks, I have seen them take out oaks)

Structure:

j'. Pl(,)(adv)P2(adv)(I')(P3)(adv)(P4)(adv)avoir(adj)(P3)(P4)(adv)p.p.(adv)
(prep)inf(adv)**P5** → **I**

    u'. P5 is the direct object → **agreement with P2**

        g'. P2 = "que" → **agreement with Pl**

        h'. if P3 exists → **agreement with Pl**

We have j' (true) u' (true) g' (false) et h' (true) → **agreement with Pl**.

We see yet again that the lexis, syntax and morphology in linguistic analysis are involved together in an organised system.

Giving syntactic norms prominence enables us to work in intension. For example we know that every French adverb can be recognised in the structure above at the positions (adv).

Later, when we review applications of our theory we will also see that one and the same model can serve for applications as different as sense extraction or 'sense mining', and machine translation.

For the moment let us see an application of our model for the agreement of the French past participle to the problem of automatic correction.

Take the sentence:

Les lettres qu'elle avait espéré leur envoyer
(The letters that she had hoped to send them)

Structure:

d'. Pl(,)(adv)P2(adv)(I')(P3)(adv)(P4)(adv)avoir(adj)(P3)(P4)(adv)**List**(adv)
(prep)(P5)inf(P5) → **I**

The agreement is correct.

Now take:

Que j'en ai vus sortir sous les hues !
(I saw so many booed on leaving!)

Structure:

a'. Pl(,)(adv)**en**(P3)(adv)(P4)(adv)avoir(adj)(P3)(P4)(adv)p.p.(adv)(prep)
(P5)inf(P5) → **I**

The agreement is incorrect, the past participle should have been written *vu* (without s).

2.2.1.5.2 *Lexical syntax.* We have already mentioned sets of lexical units (compounds). We give here an example of how syntax works in compounds.

Looking at set expressions (more or less frozen) we have discovered that French compounds with four elements could enter in (match with) 21 different structures (Echchourafi 2006). We give an extract from these structures with accompanying examples.

- V + N + (et/en/ou/pour/de/au) + N
  (constraint: the two nouns have to be common nouns and not proper nouns)

For example:

> *rendre pois pour fèves* (set expression)
> *prendre Paris pour Rome* (free lexical units)

- V + N + au + N
  (constraint: the two nouns have to be common nouns and not proper nouns)

For example:

> *avoir gain de cause* (set expression)
> *visiter Paris au printemps* (free lexical units)

We remark that in the two constraints above, which are identical:
(constraint: the two nouns have to be common nouns and not proper nouns)

the category of nouns being partitioned into proper nouns and common nouns, the constraint can be rewritten:
(constraint: the two nouns have to be common nouns)

thus avoiding an explicit negation involving an open set. This example illustrates our practice of avoiding explicit negations which here has been facilitated by our methodology which uses set partitioning. As we will explain later concerning the application of the abstract mathematical model, our practice ensures that we do not disallow constructive models.

2.2.1.6 The same formal representation over domains and languages

We have talked about canonicals and variants; we now also talk about norms and divergences.

Many Arabic utterances from the simplest to the most complex, and which contain important semantic elements can be quite simply represented for example by the structure shown in Figure 49. We do not need transformational analysis to go from an affirmation to a negation. We notice that syntax, morphology, as well as certain categories that our needs require, are all present together in this structure. The structure invokes the micro–systems needed for the particular type of problem to be processed.

opt(particule(s)) + (…) + verbe + opt(particule(s) + (…) + sujet + opt(particule(s)) + (…) + cod + opt(particule(s)) + (…)

Figure 49. Structure covering many Arabic utterances from the most simple to the most complex

The descriptive model always uses the same rule format.

In Figure 50 we show two rules where one is for machine translation (French to Arabic) and the other for sense mining (Chinese); the formal representation is the same (Cardey et al. 2010).

Machine translation:
opt(neg1) + lexis(' يجب ') + opt(neg2) + nver +arg1(acc) + opt(opt(prep_comp1),comp1(n)) + opt(opt(prep_comp2),comp2(n)) + pt

Sense mining:
l(d) + '['(_) + 从 + l(numerals) + 到 / 至 + l(numerals) + l(time) + ']'(_) + l(f)

Figure 50. Example of rules for machine translation and sense mining

For sense mining, take the following sentence:
المجموعة حضّرت الاعتداء في لبنان ← The group prepared the attack in the Lebanon.

The structure which has enabled finding this information is as follows:

opt(Part. interrog) + opt(Part. nég) + opt(Part.prob) + Primat + opt(Part.prob) + Prédicat + opt(Pronom.connect) + opt(Cod1>>Cod2>>Cod3) + opt(Prép) + opt(Coi) + opt(opt(Prép) + Ccirt) + opt(opt(Prép) + Ccircl) + opt(opt(Prép) + Cman)

With this same structure, other types of information can be found:

– Information 2 found:
العصابة سرقت المخزن بالأمس ← The gang stole from the shop yesterday

Relation_Seme_Name:	年代,year	(年代 = year)
Text Before:	""	
Matched Text:	" 从 80 到 90， 年代"	(From 80 to 90)
Text After:	"劳动是朝鲜研制的中程弹道导弹，该导弹是前苏联地对地导弹飞毛腿 C 的改良型 (射程约 500 公里), 射程 1300 公里。 (LAODONG is a medium-range ballistic missile developed by North Korea. The missile is an improved version of the Soviet Union surface -surface Scud C (range of 500 km), with a range of about 1300 km.)	
Compared_Informations:	[从,80,到,90, 年代]	(From 80 to 90)
Structure used:	1. l(d) + '['(_) + 从 + l(numerals) + 到 / 至 + l(numerals) + l(time) + ']'(_) + l(f)	

Figure 51. Example of comparing some information in a large Chinese corpus

—   Information 3 found:

المجرم قتل الشرطي في العاصمة ← The criminel killed the police officer in the capital.

We can take our model and apply it to Chinese.

The application concerns comparing some information by searching in a large corpus; see Figure 51.

With this same structure, we can also find the following Chinese sentences:

从 60 到 90 年代，   (From 60 to 90)
从 60 至 80 年代，   (From 60 to 80)
从六十至八十年，   (From 60 to 80)
etc.

Concerning sense mining, the norms and the divergences are contained in a single set of structures because the objective is to find all the different ways of saying the same thing.

In the three preceding examples (automatic correction of the agreement of the French past participle, Arabic sense mining, comparing information in Chinese) we can see that our model can be used for different applications over different languages.

2.2.1.6.1 *Interlanguage norms.* We now turn to looking at norms from one language to another.

We are going to see that some languages present similarities but that they also present numerous divergences.

The form of the sentence is not the same in different languages (Kuroda & Chao 2005). The concept of separator, gender, number, inflexion, verbal complex and its construction, classificator, serial noun, juxtaposition, coordination, common property, anaphora are different or absent according to the language. For machine translation, 'gaps' occur between the source language (SL) and the target language (TL): semantic distinctions which are finer in the TL than in the SL, and vice versa. One needs to take into account divergences such as:

- lexical density divergences (one unit in one language can correspond to several combined 'words' in the other)
- categorical divergences (a noun translated by a verb)
- syntagmatic divergences (a noun phrase in one language can be translated by a conjunctional proposition in French for example)
- syntactic construction divergences (complement order inversion for example)
- anaphoric divergences (zero for Chinese, pronoun for French)
- Determinants for example vary in their use in French in comparison with English.

2.2.1.6.2 *Divergent structures.* We give below two syntactic structures as an example; one is Arabic and the other Japanese. The two structures are completely divergent even though they express the same meaning:

- Arabic: opt(quest + يمكن +nver) + opt(ecd) + comps/compscc/compa/compacc + point
- Japanese: comps/compscc/compa/com-pacc + opt(ecd + opt(quest + vinf)) + point

On the contrary, one observes that the two following structures (one Arabic, the other English) are identical:

- Arabic: opt(neg) + pred + opt(mod) + opt(compacc/compa1) + opt(compacc/compa2) + opt(compacc/compa3)+ opt(mod) + opt(2pts+ liste) + point
- English: opt(neg) + pred + opt(mod) + opt(compacc/compa1) + opt(compacc/compa2) + opt(compacc/compa3) + opt(mod) + opt(2pts + liste) + point

In one of the applications that we will see later in Part 3, that of machine translation, we have compared the norms of each of the languages that we process and kept those which were going to enable us to obtain the best translations.

For example, for the French:

> *Réduire la vitesse en dessous de 205/.55.*
> (Reduce the speed below 205/.55.)

we have the structure:

> frC_7
> opt(neg1) + opt(neg2) + vinf + arg1 + prep_v + arg2 +
> opt(opt(prep_comp), comp1(n)) + opt(opt(prep_comp),comp2(n)) + pt

and the corresponding Arabic sentence:

يجب تقليص السرعة تحت 205./55

with its structure:

> arC_7.a
> opt(neg1) + lexis(' يجب ') + opt(neg2) + nver + arg1(acc) + prep_v +
> arg2(acc) + opt(opt(prep_comp1),comp1(n)) +
> opt(opt(prep_comp2),comp2(n)) + pt

For the French sentence:

> *Signaler le cas à l'Institut de Veille Sanitaire immédiatement.*
> (Report the case to the Institute for Health Surveillance immediately.)

we have the same structure as before:

> *frC_7*
> opt(neg1) + opt(neg2) + vinf + arg1 + prep_v + arg2 +
> opt(opt(prep_comp), comp1(n)) + opt(opt(prep_comp),comp2(n)) + pt

However for the Arabic translation:

يجب إعلام مركز المراقبة الصحية بالحالة فورا

we have a different structure:

> arC_7.b
> opt(neg1) + lexis(' يجب ') + opt(neg2) + nver + arg2(acc) + prep_v +
> arg1(acc) + opt(opt(prep_comp1),comp1(n)) +
> opt(opt(prep_comp2),comp2(n)) + pt

Thus we note, by means of these two examples, that the same French structure *frC_7* which refers to the two French verbs *réduire* and *signaler* gives two different Arabic structures *arC_7a* and *arC_7b*. If the structure of the first French verb *réduire* is nearly identical to that of the Arabic verb, the same structure *frC_7* of the verb *signaler* necessitates a permutation of the two arguments arg1 and arg2 in Arabic, and this gives the Arabic structure *arC_7b* which is totally different to that of the French.

In examining Figure 52, one can see some problems of interlanguage ambiguity (Cardey 2008b). We give here the example of the translation of prepositions and the problems they can pose. In our applications they are controlled according to their use.

Language	Preposition (French)	Possible translations	Examples (in French)
Arabic	*à*	فى	*je vis à Toronto.* (I live in Toronto)
		على	*je te verrai à six heures* (I will see you at six o'clock)
		ل	*Ce jouet est à Gilles* (This is Gilles' toy)
		Ø	*une cuillère à soupe* (a soup spoon)
		ب	*des livres à dix dollars* (books at six dollars)
Chinese	*par*	从	*par la fenêtre* (by the window) *par la poste* (by post)
		每	*une fois par an* (once a year)
		Ø	*par exemple* (for example)
		用	*prendre par la main* (take by the hand)
Japanese	*sur*	にかけた	*faire attention à l'huile sur le feu et aux grille-pains* (be careful with oil on the burner and toaster)
		の上の	*objets sur l' étagère* (objects on the shelf)
		に	*ne pas verser d'eau sur de l'huile en feu* (do not pour water on burning oil)
		を	*ramper sur le sol* (crawl on the floor)
Thai	*par*	ด้วย	*remplacer qqn par qqn* (replace someone with someone)
		จบด้วย	*finir par* (finish by)
		ผ่านทาง	*passer par* (pass by)
		ทาง	*par courrier* (by mail)
		ต่อ	*une fois par an* (once a year)

Figure 52. Preposition use in French to Arabic, Chinese, Japanese and Thai

The following French example highlights the problem:

> *Nous avons convenu avec le secrétaire de l'agence de voyage de l'heure de départ de l'autobus.*
> (We have agreed the bus departure time with the travel agency secretary.)

Which *de* attaches to *convenir*? For machine translation, this is far from being simple and therefore either this type of construction ought to be avoided or it ought to be controlled.

We have the same situation for the following type of ambiguity:

> *Elle ne rit pas parce qu'il est idiot.*
> (She does not laugh because he is an idiot.)

where there are two interpretations; in one case *elle rit* (she laughs) and in the other case *elle ne rit pas* (she does not laugh).

Given a defined domain and a specific need to be processed, one can thus represent the interferences and divergences between the languages concerned in the following way.

Let us take three (or more) languages, whatever they are. We have as a whole the systems which are common to the three languages in question and from which it would suffice to take, and to which would be added the specifics of two of the languages and then of each of the languages during the process of automatic translation, see Figure 53.

It is already possible to imagine that certain of the systems will be common with inflexional languages, others with agglutinative languages and still others with isolating languages.

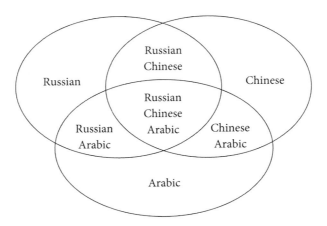

Figure 53. Common and specific systems between languages

As we have said at the outset, our model can be applied to all languages as well as modelling and representing whatever we wish to bring to the fore or to give prominence to, whether in the same language or between languages.

2.2.1.7   Disambiguation

To conclude this section concerning the linguistic model, we give three examples of disambiguation using the model.

2.2.1.7.1   *Disambiguation in English.* Given the need of disambiguation in English we will see that this necessitates partitioning the English lexis in parts of speech.

We determine the ambiguous classes in the sense that these contain not only the ambiguous forms but also the non ambiguous forms. Referring to Figure 54, (Birocheau 2004: 68), for example the class of adjectives (ADJECTIVES) contains the class nouns/adjectives (Noun/Adj), this latter also being an element of the class nouns (NOUNS). Noun/Adj/Verb is an element of Noun/Verb, of Noun/Adj, of ADJECTIVES, of VERBS and of NOUNS.

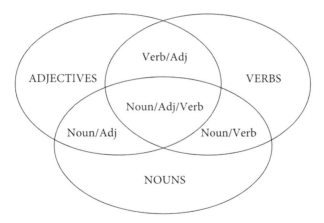

Figure 54. Class ambiguities

This same model enables showing that for French there are 29 classes of ambiguity.

Let us now look at two isolating languages, Thai and Chinese.

2.2.1.7.2   *How to represent a problem of ambiguity in composition: Thai.* For Thai, as well as tonemes which can affect the meaning of a word, the lack of visible frontiers between syntagms can pose problems of ambiguity when translating Thai to another language (Kiattibutra 2011). The written language has no spaces between the words in a sentence, and the sentence length increases the difficulty of toneme and grapheme recognition; this is why it is difficult to identify these correctly in an automatic manner. If this identification is done incorrectly, the words which are at the frontier between two syntagms can be included in false compounds as the following examples show.

Take the sentence:

[ผู้นำเสนอข้อต่อรองราคาข้อต่อทีเขาต้องการซื้อจำนวน 3 ข้อต่อคณะกรรมการ]

whose translation into French is:

[Le chef propose une négoci ation du prix du joint qu'il veut acheter au-près du comité sous 3 conditions.]
(The manager proposes a negotiation of the price of the joint that he wants to buy with the committee subject to 3 conditions.)

Figure 55 shows an analysis of the example sentence.

The analysis shown in Figure 55 reveals that there are in fact three analyses possible, identified as A, B and C.

	1	2	3	4	5	6	7	8
A	-	ผู้นำ (n)	เสนอ (v)	-	ข้อต่อรอง (n)	-	ราคา (n)	-
	Le	Chef	propose	une	négociation	du	prix	du
B	ผู้ (n)	นำเสนอ	ข้อต่อ (n)	รอง (v)		-	ราคา (n)	-
	Homme	Présent	joint	appuyer		du	prix	du
C	ผู้ (n)	นำเสนอ	ข้อ (n)	ต่อรอง (v)		-	ราคา (n)	-
	Homme	Présent	condition	négocier		du	prix	du

	9	10	11	12	13	14	15	16
A	ข้อต่อ (n)	ทีเขา (pron)	ต้องการ (v)	ซื้อ (v)	จำนวน 3 ข้อ (adj+cl)	ต่อ (prép)	คณะกรรมการ (n)	/-/
	joint	qu'il	Vent	Acheter	sous 3 condition	auprès du	comité	/./
B	ข้อต่อ (n)	ทีเขา (pron)	ต้อง (v)	การซื้อ (n)	จำนวน (adj)	ข้อต่อ (n)	คณะกรรมการ (n)	/-/
	joint	qu'il	Doit	Acheter	sous 3	Joint	comité	/./
C	ข้อต่อ (n)	ทีเขา (pron)	ต้อง (v)	การซื้อ (n)	จำนวน (adj)	ข้อต่อ (n)	คณะกรรมการ (n)	/-/
	joint	qu'il	Doit	Acheter	sous 3	Joint	comité	

Figure 55. Examples of false compounds in Thai.

For example, we extract from the sentence:

ข้อต่อรอง

The structural order of the Thai sentence is not the same as that of French. The word [ข้อต่อรอง] is a compound noun of type [noun + verb + noun] which can be translated into French by *négociation* (cell A5 in Figure 55). However if we extract the word [ข้อต่อ] *un joint* (cell B3), we can also obtain the verb [รอง] *appuyer* (cell B4), and so on.

Figure 56 illustrates the potential for extraction from ข้อต่อรอง.

The various possibilities are marked in a corresponding fashion in Figures 55 and 56.

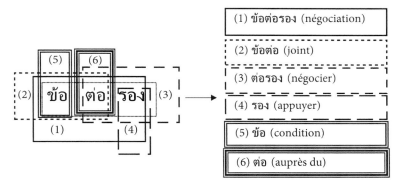

Figure 56. Illustration of the potential for extraction from ข้อต่อรอง

We will see how to solve this problem in Thai with the following example in Chinese, where the method for Thai is the same as for Chinese.

2.2.1.7.3 *Disambiguating Chinese and Thai.* For Chinese the problem concerns grammatical category (parts of speech) ambiguity, the example being due to Gan Jin.

We take for example:

好(N, Adj, Adv, V)

where N is variously noun, Adj adjective, Adv adverb and V verb.

a. The case of nouns N: V + 好+ opt(儿)   (well, good, "good day")
    – 这场戏，观众连声叫好。   The audience applauded the play several times.
    – 如果你去见老师，给我捎个好儿。   Say "good day" when you see the teacher.

b. The case of adjectives Adj: (excellent quality, satisfaction)
    1. 好 + opt(的) + N
        –   好人。   Good person.
        –   好的歌曲。   Good song.
    2. N + opt(Adv) + 好 + opt(了)
        这本书很好。   It is a very good book.
        –   质量明显的好了起来。   The quality has considerably improved.
    3. opt(N) + 好在 + opt(N) + V   (the reason for which)
        他好在对人诚恳。   Happily, he is sincere with others.
    4. 还是 + VG + opt(的) + 好   (in comparison, I would prefer (verbal group VG))
        –   你还是别去的好。   It would be better that you do not go.
    5. N + opt(V) + 好 + 了   (end of some action)
        –   饺子(做)好了。   The ravioli is ready.

c. The case of adverbs Adv:
    1. 好 + opt(不) + Adj
        –   市场上好(不)热闹。   The market is very lively.
        In this example, semantically 不 is a negation, but here 好不热闹 (lively) = 好热闹(lively). This is so even with an exception with 容易 (easy). 好不容易 (not easy) = 好容易(not easy) (Shuxiang 1979: 175).
    2   N + 好 + V
        –   我们好找了一通。   We have really searched hard.

d. The case of auxiliary verbs V:
        N + 好 + V
        –   带伞，下雨好用。   Take the umbrella; you can use it if it rains.

These various structures serve uniquely for segmenting a Chinese sentence.

These examples can be represented as in Figure 57 (due to Gan Jin).

The same model representation can be used for our preceding problem with the Thai language.

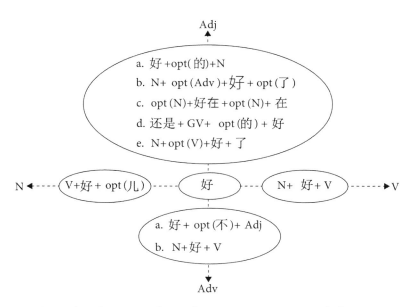

Figure 57. Examples of a system of sets of structures serving uniquely for segmenting a Chinese sentence

2.2.2 The mathematical model

Again, let us draw close to our galaxy and to the stars. Our system, our systems are in a state of disequilibrium because they exchange with their environment. There is thus a situation of stability within a real disequilibrium due to external flows.

We shall see this in detail when we describe the canonical elements and their variations according to the context. The invariant here is due to the rules' stability; we have seen that the rules, that is, grammar, almost never change in synchrony, because if this were so, it would be the whole system which would have to be reconsidered. This means too that the rules' organisation also is stable despite the flows that traverse them. The same equilibrium is assured for the whole of the macro-system thanks to the actions and retroactive effects which keep the whole stable as with the galaxies' stars.

We use set theory and in particular partitioning as the basis of our model for decomposing the material making up language.

2.2.2.1 Necessary notions

A brief reminder concerning set theory, partitions, logical operations, and relations.

It is to De Morgan (1806–1876) and above all Boole (1815–1864) to whom one has to go back for the effective foundation of an algebra of "operations of thought".

2.2.2.1.1 *Set*

What we need to know for our model.

A set consists in the juxtaposition of several elements (constituents) which are objects. An object is a set which is at most unary, that is, a set comprising at most one and only one (unique) element.

Logic imposes on sets certain experimentally verifiable conditions:

1. It is possible to isolate their elements which should not have any common parts;
2. it is possible to determine the number of their elements;
3. it is possible to decide without ambiguity to which set an object belongs.

In set logic, the sentence:

"The sun is a star"

is expressed by:

(the object) sun is an element of the set of stars

Two different writings of sets differing only in the order in which the elements are named designate the same set.

A written 'word' can be a set of letters or a set of symbols; as such it is an ordered set, the order being significant.

In French:

rame ≠ amer ≠ mare ≠ arme

but:

$\{r,a,m,e\} = \{a,m,e,r\} = \{m,a,r,e\} = \{a,r,m,e\}$

Equivalence

There is a semantic equivalence between two named objects if and only if they designate the same object. Thus:

$2 + 3 = 5$

That which designates 2 plus 3 and that which designates 5 is the same thing.

When we write a = b, we give two different names to one and the same thing.

Equality

Leibniz' law: two things are equal if and only if every property of one is the property of the other and vice versa.

Two equivalent things from one point of view may not be from another point of view.

Example: two disyllabic words are equivalent in respect of their length measured in syllables but not necessarily for the rest.

Two sets of the same cardinality (number of elements) are said to be equipotent. There is a one–to–one (biunivocal) correspondence between the elements of equipotent sets.

Every set comprising two elements is a pair.

A set is singleton if and only if it comprises one and only one element, if and only if its cardinality is one.

We call empty set the set which comprises no elements (and of which the cardinality is zero).

We write the empty set symbolically as $\varnothing$.

Set hierarchy

We can have an alphabet as a set of letters.

We can form the set of the Greek, Cyrillic, Latin and Scandinavian alphabets. Let us call such a set of sets E.

We can write the Greek alphabet $\in$ E.

A set can be an element in respect of another set.

We can say that the letter $\alpha \in$ the Greek alphabet $\in$ E.

$\alpha \notin$ (is not an element of) E because α is not an alphabet.

Four operations on sets

Intersection

The intersection of the set A by the set B is the set formed by the elements of A which are present in B.

We write this as $A \cap B$.

Union

The union of the set A and the set B is the set formed by the elements of A to which are added the elements of B which are not elements of A.

We write this as $A \cup B$.

Difference

The difference of the set A and the set B is the set formed by the elements of A to which one has removed the elements of A which are also elements of B.

We write this as $A \setminus B$.

$A \setminus B$ is also called the relative complement of B in A.

Prolongation

To pass from intersection to union, it is necessary to perform the union of the intersection and the parts that are not in common.

The prolongation of the set A by the set B is the union of their parts that are not common. We can also say that the prolongation is the complement of the intersection with respect to the union.

We write this as A Δ B.

Thus:

$$A \Delta B = (A \setminus B) \cup (B \setminus A)$$

Complement

The complement of a set consists in subtracting it from the reference domain. The new set thus obtained is called the complement of the set.

As an aid to understanding, try the following exercise (translated from Gentilhomme (1974)):

> *Zizi's misfortunes*
> "Dad told me like so:
> 'Zizi, I want you to be a good daughter, so you know how to get on in the world. You've again got a zero in spelling. So listen to me, for punishment and to learn you to like good literature, this Sunday afternoon you're to copy me all the names which are at the same time in act 1, scene 2 of ATHALIE, in act 2, scene 3 of BERENICE and in act 3, scene 1 of CLITANDRE. You're to add to this list all the names in act 1, scene 2 of ATHALIE after you've removed all those in act 3, scene 1 of CLITANDRE and those in act 2, scene 3 of BERENICE. You're following me? Then you're to add to what you've done all the names which are at the same time in act 1, scene 2 of ATHALIE and in act 3, scene 1 of CLITANDRE from which you'll have already crossed out all the words in act 2, scene 3 of BERENICE. Then, and don't forget, you're to add to all this the same thing but with CLITANDRE replaced by BERENICE and BERENICE by CLITANDRE.
> Finally, and be careful, in what you've written you're to erase all the words which are in the same act and scene of ATHALIE I've just said. And make certain it's done neatly; otherwise you're in for a hiding. When you'll be a grown-up you'll thank me for all I've done for you.'
> I gave Dad a blank sheet of paper, and I still got a hiding. It's not fair!"
> What do you think?

The solution is given in Figure 58.

Data: Relations between sets expressed in natural language
Processing: Translate into 'set language'. Perform the calculation.
Let:

      A be the set of all the names in act 1, scene 2 of ATHALIE,
      B be the set of all the names in act 2, scene 3 of BERENICE,
      C be the set of all the names in act 3, scene 1 of CLITANDRE.

We thus have:

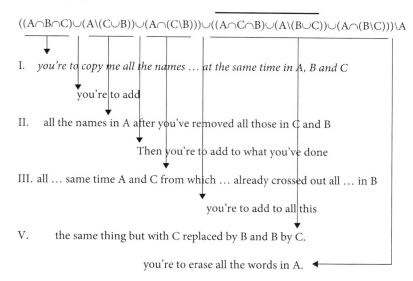

$$((A \cap B \cap C) \cup (A \setminus (C \cup B)) \cup (A \cap (C \setminus B))) \cup ((A \cap C \cap B) \cup (A \setminus (B \cup C)) \cup (A \cap (B \setminus C))) \setminus A$$

I.   *you're to copy me all the names … at the same time in A, B and C*

        you're to add

II.   all the names in A after you've removed all those in C and B

        Then you're to add to what you've done

III. all … same time A and C from which … already crossed out all … in B

        you're to add to all this

V.   the same thing but with C replaced by B and B by C.

        you're to erase all the words in A.

where (IV) = (I) $\cup$ (II) $\cup$ (III).
We have:

      (I) $\cup$ (II) $\cup$ (III) $\cup$ (IV) $\cup$ (V) = A

Finally,

      A \ A = $\varnothing$.

However the text is ambiguous at two places:

1. It is not clear if we have A\(C$\cup$B) or (A\C)\B), but this does not matter
    as the two results are the same.
2. One does not know if the part (A$\cap$C$\cap$B)$\cup$(A\(B$\cup$C)) should or should
    not be considered.

Whichever way we interpret the ambiguities, we get the same result, the
empty set.

Conclusion: Zizi was right, the punishment was unjustified!

Figure 58. Solution to Zizi's problem

2.2.2.1.2 *Partition*

The concept of the partition of a set E is a response to the idea of having an ideal
classification of a list of objects in distinct lots, called classes. More precisely, we
have three imperatives:

1   No class is empty; that is to say that each lot should contain at least one object;
2   The classes are disjoint; that is to say that the lots are separate one from the other;
3   The union of the classes returns the set E; that is to say that all the objects of E have been allotted and that none of them has been forgotten during the operation.

If we have 3 classes A, B, C, the 3 imperatives above can be written symbolically as follows:

1    $A \neq \varnothing, B \neq \varnothing, C \neq \varnothing$
2    $A \cap B = A \cap C = B \cap C = \varnothing$
3    $A \cup B \cup C = E$

We can formulate numerous partitions each time adopting some particular reason for considering objects equivalent to others.

For example the traditional sense of the term part of speech to which one can associate a 'word' or rather a lexical item is sufficient for obtaining a partition of the set E but not of the set F, the reference domain being French where E and F are thus:

E = {jour, chantait, cheval, beaucoup, chanson, cœur, faiblement, très},
F = {ferme, chasse, couvent, sous, grave}

this being due to the fact that each of the elements of set F, unlike E, is an element of more than one part of speech.

For a given partition, each class is entirely determined by the particular reason pertaining to one of its representatives. The object chosen for representing its class is called the 'canonical' element.

Let us now look at crossing partitions and this by means of an exercise; see Figure 59.

Subsets
Given two sets S and T, if all the elements of T are also elements of S, we write:

$T \subseteq S$ (T is a subset of S),

and

$S \supseteq T$ (S is a superset of T).

If $T \subseteq S$ and $T \neq S$, we write:

$T \subset S$ (T is a proper subset of S),

and in like manner:

$S \supset T$ (S is a proper superset of T).

Let E be the set:

{chanter, robe, livre, écrire, chapeau, carnet, violon, chaussure, pianissimo, cahier, flûte, guitare, veste, violoncelle, pantalon, chausser, lire, moderato, habiller, allegro, crayon, noter, livresque, apprendre, manteau}

Construct 4 partitions P_1, P_2, P_3 and P_4 over E according to the following criteria:

a: length of words as number of letters
b: nature of the first letter of the word (e.g. vowel, plosive, etc.)
c: part of speech to which the word belongs:
     {verb, noun, adjective, adverb}
d: semantic field: {music, clothes, studying}

Cross the 4 partitions, 2 by 2, then 3 by 3, then all the 4. Classify the resulting partitions according to their degree of refinement.

Solution.
We have:

$$P_1 \cap P_2 = P_2 \cap P_1 \qquad \text{(commutativity)}$$
$$P_1 \cap P_2 \cap P_3 = (P_1 \cap P_2) \cap P_3 = P_1 \cap (P_2 \cap P_3) \quad \text{(associativity)}$$

Crossing 2 by 2 we obtain 6 possible partitions which are more refined than the 4 constructed at the beginning of the exercise.

For 3 by 3 we obtain:
    abc, abd, acd, bcd
which are even more refined.

abcd is the most refined.

Figure 59. Exercise on partitions

2.2.2.1.3   *Logical operations*

Let P and Q be propositions.

A proposition with the value TRUE is indicated by 1.

A proposition with the value FALSE is indicated by 0.

The logical operations and their operators that we will use are given in Figure 60.

2.2.2.1.4   *Binary relations*

The system which links a canonical element to its variants is based on binary relations. We provide here a brief review of binary relations based on (Lopota 1981), followed with observations concerning their properties and use for our purposes.

The binary relation $\mathcal{R}$ is defined by its graph $\mathcal{G}$, this being a subset of the Cartesian product of a couple of sets, D (departure) and A (arrival) (A can be equal to D).

$\neg$ **Negation**

P	¬P
1	0
0	1

Remark : the notation $\bar{P}$ is equivalent to ¬P.

$\vee$ **Disjunction**

P	Q	P ∨ Q
1	1	1
1	0	1
0	1	1
0	0	0

$\wedge$ **Conjunction**

P	Q	P ∧ Q
1	1	1
1	0	0
0	1	0
0	0	0

$\Rightarrow$ **Implication**

P	Q	P ⇒ Q
1	1	1
1	0	0
0	1	1
0	0	1

$\Leftrightarrow$ **Bi-implication or bicondition**

P	Q	P ⇔ Q
1	1	1
1	0	0
0	1	0
0	0	1

Figure 60. Logical operations and operators

The Cartesian product of D by A, D × A, is the set of couples with 1st element in D and 2nd element in A.

We have:

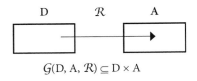

$$\mathcal{G}(D, A, \mathcal{R}) \subseteq D \times A$$

The domain of definition of $\mathcal{R}$ in D, $\mathcal{D}$, is the set of elements of D having at least one corresponding element in A.

The codomain (image) of $\mathcal{R}$ in A, $\mathcal{C}$, is the set of elements in A corresponding to an element in D. This set can be a singleton or indeed empty.

The principal properties of binary relations are as follows:

- $\mathcal{F}$ to be a *function*: the image of an element x ∈ D contains at most one element.
- $\mathcal{A}$ to be an *application* (or total function): an application is a function for which all the elements of D have one and only one corresponding element in A; that is $\mathcal{D} = $ D.

– *I* to be *injective*: a relation is injective if 2 distinct elements in D never have the same corresponding element in A.
– *S* to be *surjective* (or onto): all the elements of A have at least one corresponding element in D.

With these properties (presence or absence of variously $\mathcal{F}$, $\mathcal{A}$, I, S) we obtain a classification of 12 types of binary relation, as shown in Figure 61, and which involves a particular algorithmic instance.

We make the following observations concerning this classification of binary relations.

A binary relation that is an application is necessarily a function but the converse does not hold. This classification does not admit an empty codomain (image); that is, any empty binary relation.

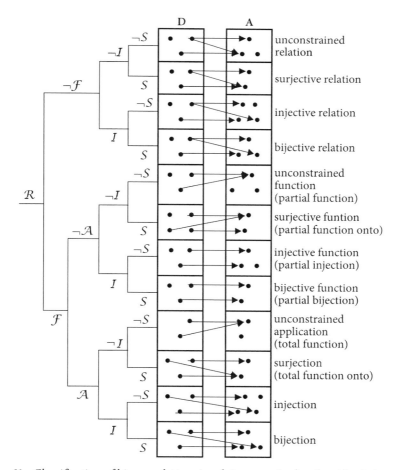

Figure 61. Classification of binary relations involving a particular algorithmic instance

The classification of binary relations in Figure 61 is in fact an algorithmic micro–system. The access to the values of the properties $\mathcal{F}$, $\mathcal{A}$, I, S, which are the conditions, and to the 12 types of binary relation, which are by the operators, is by means of a binary tree rooted by a unary tree with node value $\mathcal{R}$, which acts as an entry condition. This tree, albeit written left to right rather than top to bottom, is the representation of the algorithm. The algorithm shown is not unique.

2.2.2.1.5 *Modelling algorithmic operations*

As we have already seen with our algorithms, we use logical implication for our operations:

$$P \Rightarrow Q$$

where P is called the conditional part and Q the conclusion part or action part, which for us are respectively condition and operation. Once a situation is recognised, the action is executed. Each time P has the value TRUE, one can deduce Q by the rule of Modus Ponens (if $P \Rightarrow Q$ is TRUE, and if P is TRUE, then necessarily Q is TRUE; this rule expresses in fact the first line of the truth table for logical implication; see Figure 60).

2.2.2.2 Mathematical modelling of micro-systemic linguistics

We now develop the abstract mathematical model of systemic and micro-systemic linguistics (Greenfield 1997), (Greenfield 2003), (Cardey & Greenfield 2005). This involves the core of systemic linguistics which itself involves the system of canonicals and variants.

Systemic linguistics methodology consists in analysing a linguistic system in component systems as follows:

- Sc: a system which is recognisably canonical;
- Sv: another system representing the variants;
- Ss: a 'super' system which puts the two systems Sc and Sv in relation with each other.

We will go back to our example of the doubling or not of the final consonant in English words before the endings -ed, -ing, -er, -est, -en.

For example we observe the variants 'modeling' and 'modelling' for the canonical form 'model'. The system, which we call *Doubling_or_not*, comprises the following component systems:

- $Sc_{Doubling\_or\_not}$, the words concerned in their basic form, that is their canonical form; e.g. 'model', 'frolic';

– $Sv_{Doubling\_or\_not}$, the words concerned in their derived or inflected form, the variants; e.g. 'modeling', 'modelling', 'frolicked';

– $Ss_{Doubling\_or\_not}$, the super-system relating systems $Sc_{Doubling\_or\_not}$ and $Sv_{Doubling\_or\_not}$.

2.2.2.2.1 *Establishing a systemic linguistic analysis*

This requires modelling system Ss. To do so for some application, the linguist establishes two categorisations:

1. Firstly a 'non-contextual' (nc) categorisation of the canonical forms in relation with the variant forms in isolation, the context being limited to just the canonical and variant forms themselves. For *Doubling_or_not*, this categorisation can thus only depend on the form of the words concerned, this being an aspect of morphology.
2. Secondly an 'in-context' (ic) categorisation of the canonical forms in relation with the variant forms in terms of the linguistic contexts of the variant forms. The systemic analysis reveals precisely which other internally related linguistic systems are involved.

For systems Ss, Sv and Sc, let S be a set structure modelling super-system Ss, let C be the set of canonical forms, V the set of variant forms, and let CV be the binary relation between C and V corresponding to system Ss.

Each of the above two categorisations, 'nc' and 'ic', can be modelled by a partition on CV; we have P_{nc} and P_{ic}. Given that we have partitions, from the fundamental theorem on equivalence relations, it follows that there exist two corresponding equivalence relations E_{nc} and E_{ic} on CV. Each equivalence class in respectively E_{nc} and E_{ic} corresponds to a distinct categorisation or case. We model system Ss, the super-system relating systems Sc and Sv, by means of the binary relation S between the equivalence relations E_{nc} and E_{ic}, and similarly S^{-1} between E_{ic} and E_{nc}.

We can subsequently model functions for finding the canonical element(s) corresponding to a variant element and vice-versa or others such as finding the (name of the) canonical equivalence class for a variant element as for example in parts of speech tagging. Furthermore, because we have a precise structure for S, we can verify that a given linguistic analysis representation is well formed. In respect of equivalence relations and when the linguistic domain consists of strings, we note that every finite automaton induces a right invariant equivalence relation on its input strings (Hopcroft & Ullman: 1969: 28–30).

2.2.2.2.2 *Establishing the partitions*

We now turn to how to establish the partitions P_{nc} and P_{ic}.

Here the linguist can adopt either a proof theoretic or a model theoretic approach–or indeed combine these (Greenfield 1997). The former is allied to the development of an algorithm, the latter being case (truth table) based.

In any case, if the goal is an automated application, an eventual algorithm is typically necessary; see for example (Humby 1973, Dial 1970).

We illustrate an algorithmic analysis with *Doubling_or_not*. Using a binary divide and conquer approach the linguist determines an algorithm for each of P_{nc} and P_{ic}; many equivalent forms of linguistic formalisation representations have been devised, and for many of these computerised source representations have been implemented, including automated transcoding between the latter (Cardey & Greenfield: 1992). As to which partition P_{nc} or P_{ic} to start with and even whether it is possible or feasible to sequence the establishment of the two partitions depends on various factors such as:

–   What prior knowledge is available. For example existing classifications as for example a parts of speech tag set;
–   The simplicity or otherwise of organising observations including their extraction. For example in machine translation and in concept mining, concepts which will constitute the canonical forms are themselves often revealed during the analysis process at the same time as the contexts indicating their presence as variants in the language.

Whatever approach is adopted, it is incumbent upon the linguist to provide a justification for each equivalence class so discerned, for both the non-contextual and the in-context analysis. Such justifications normally consist of an observed, that is attested, canonical form, its associated variant form, and the provenance of this observation. The couple:

(canonical from, associated variant form)

is the representative element of the equivalence class.

2.2.2.2.2.1 *The non-contextual analysis.* Figure 62 shows a representation of the result of a non-contextual (nc) analysis carried out by the linguist for Doubling_or_not.

The conditions and the operators are abstraction predicates over *CV*. The algorithm's entry condition identifier 'Id' is cv, the abstraction predicate for the set *CV*, this being the linguistic domain under analysis. The algorithm is represented in our organigram form which is well suited to the nested partitioning process carried out by the linguist during the analysis. In this respect, we remark (see Figure 62) for example the visual cues and the explicitation of the nesting level.

Doubling_or_not non-contextual (nc) analysis			
Conditions			
Id	Condition text		
cv	word with final consonant in English taking -ed,-ing,-er,-est, -en		
cvd	Doubling of the final consonant		
k	The words terminating in -ic take -ck		
Operators			
Id	Operator text		
N	No doubling of the consonant		
D	Doubling of the consonant		
K	The words terminating in -ic take -ck		
Algorithm represented as organigram with justifications			
Line #	Level	Condition -> Operator	Canonical -> Variant : Provenance
0	0	cv -> N	'feel->'feeling : CEOED 1971
1	1	cvd -> D	'run->'runner : CEOED 1971
2	2	k -> K	'frolic->'frolicked : CEOED 1971

Figure 62. Representation of the non-contextual (nc) analysis for *Doubling_or_not*

In *Doubling_or_not*, the apostrophes with the word forms are English stress indicators.

Being a partition, with each equivalence class being associated directly with a line in the algorithm allows us to include a justification for each class, that is, categorisation or case; these are shown in Figures 62 and 68 in the column **Canonical → Variant: Provenance** where for *Doubling_or_not* the provenances are variously:

– CEOED 1971:   Concise Edition of the Oxford English Dictionary, 1971
– WNCD 1981:   Webster's New College Dictionary, 1981

Apart from experimental scientific practice considerations, in the engineering perspective including such case justifications assists application evaluation processes such as validation and is typically mandatory in safety critical applications. In automated applications, automated case based regression validation testing can be implemented. Such case justifications, precisely because they are case based, can serve as the basis for exhaustive evaluation benchmarks.

The algorithm is shown in Figure 63 in conventional 'if then (else) fi' representation.

In Figure 63 it is to be observed that the 'ifs' other than the outermost are of the type 'if then else fi'. This is because in general we require that systemic linguistic analyses be exhaustive but not over-generative; the algorithm covers exactly the linguistic domain *CV* under analysis. This 'if then (else) fi' structure is equivalent to a binary tree rooted by a unary tree; hence as the algorithm's organigram representation in Figure 62 shows, the number of lines equals the number of nodes (condition appearances) equals the number of leaves (operator appearances).

if condition cv is true

   then   if condition cvd is true

            then   if condition k is true

                    then operator K

                    else operator D

                    fi

            else   operator N

            fi

   fi

Figure 63. Conventional representation of algorithm (nc)

We can write the model theoretic model of the non-contextual (nc) analysis for *Doubling_or_not* with, for convenience, each interpretation in the same order as in the organigram representation of the algorithm. In the model theoretic model (Figure 64) the conditions and operators are predicates where the items in bold correspond to those in the associated algorithm line. We formulate the model theoretic model as a single proposition; hence the disjunction of the interpretations, and thus the model theoretic model is represented in disjunctive normal form.

0. **cv** $\wedge \neg$ cvd $\wedge$ **N** $\wedge \neg$ D $\wedge \neg$ K $\vee$

1. cv $\wedge$ **cvd** $\wedge \neg$ k $\wedge$ $\neg$ N $\wedge$ **D** $\wedge \neg$ K $\vee$

2. cv $\wedge$ cvd $\wedge$ **k** $\wedge$ $\neg$ N $\wedge \neg$ D $\wedge$ **K**

Figure 64. Model theoretic model of the non-contextual (nc) analysis for *Doubling_or_not*

In each of the interpretations we observe that only one operator is positive. However, there exist linguistic systems with more than one variant for a given interpretation of the conditions; for example for the system of the plural of the French adjective, the canonical (singular) form 'austral' has variants (plurals) 'australs' and 'austraux'. Thus for the 'austral' example there are two interpretations in the model theoretic model but with condition formulae that are equal.

Concerning the algorithm and its model theoretic model, we note that the model can be generated from the algorithm. Furthermore, in general there are many functionally identical algorithms that can be generated from a model theoretic model (the conditions and operators resting unchanged). Let P_N be the number of algorithms that can be generated from a model theoretic model with N conditions, where all the 2^N possible interpretations are present. We have:

$$P_N = N \times (P_N - 1)^2$$

where $P_2 = 2$ (Humby 1973: 32–34). In consequence alternative functionally identical algorithms can be generated to meet specific needs such as speed optimisation in automated applications.

Let X and Y be sets and their predicates of abstraction be x and y respectively. We have:

- x ∧ y   corresponds to: $X \cap Y$
- x ∧ ¬ y corresponds to: $X \setminus Y$

From the model theoretic model, set expressions can thus be formulated corresponding to the conditions component of each interpretation as shown in Figure 65. Here, the set abstractions of the conditions are in italic capitals, and a Dewey Decimal based notation is used to identify each interpretation's conditions component.

Set formulation of conditions components (nc)			
Algorithm		Set name	Set formulation
Line #	Level		
0.	0	CVnc.0	$CV \setminus CVD$
1.	1	CVnc.0.0	$CV \cap CVD \setminus K$
2.	2	CVnc.0.0.0	$CV \cap CVD \cap K$

Figure 65. Set formulation of conditions components (nc)

The sets $CVnc.0$, $CVnc.0.0$ and $CVnc.0.0.0$ partition the set CV. Concerning partitioning, let X, Y, Z be sets; partition is defined as:

$$\{X, Y\} \text{ partition } Z \Leftrightarrow (X \cap Y = \varnothing) \wedge (X \cup Y = Z).$$

We observe:

- The intersection of the sets $CVnc.0$, $CVnc.0.0$ and $CVnc.0.0.0$ is the empty set:
  $\cap\{CVnc.0, Cnc.0.0, CVnc.0.0.0\} = \varnothing$
- The union of the sets $CVnc.0$, $CVnc.0.0$ and $CVnc.0.0.0$ is the set CV:
  $\cup\{CVnc.0, CVnc.0.0, CVnc.0.0.0\} = CV$

Thus $Pnc = \{CVnc.0, CVnc.0.0, CVnc.0.0.0\}$. Being a partition, the algorithm has determined an equivalence relation Enc over CV; each of the sets $CVnc.0$, $CVnc.0.0$ and $CVnc.0.0.0$ is an equivalence class. The number of equivalence classes, that is the index of the equivalence relation E_{nc}, is 3, that is, the cardinality of P_{nc}, is determined by the algorithm, and thus is also equal to the number of lines in the algorithm, and furthermore is equal to the number of operators:

$$\#P_{nc} = 3$$

Now consider the sets that are defined during the execution of the non-contextual algorithm (Figure 66).

Sets defined during execution of the algorithm (nc)			
Algorithm		Set name	Set formulation
Line #	Level		
0.	0	$CV'nc.0$	CV
1.	1	$CV'nc.0.0$	$CV \cap CVD$
2.	2	$CV'nc.0.0.0$	$CV \setminus CVD \cap K$

Figure 66. Sets that are defined during the execution of the algorithm (nc)

Proper subset structure (nc)	
Algorithm line #	Parent set $\supset$ Line set
1.	$CV'.0 \supset CV'.0.0$
2.	$CV'.0.0 \supset CV'.0.0.0$

Figure 67. Proper subset structure (nc)

The sets so defined form a collection of proper subsets (Figure 67).

To summarise in respect of the non-contextual analysis, this results in a partition P_{nc} over CV, each element of P_{nc} being an equivalence class in which each canonical element shares the same way of forming its related variant element independently of the context ('nc' implying 'no context'). The index of the equivalence relation $\#P_{nc}$ is equal to the number of operators, 3 for the *Doubling_or_not* example.

2.2.2.2.2.2 *The in-context analysis.* The in–context (ic) analysis carried out by the linguist follows the same approach as the non–contextual analysis. However, the non–contextual (nc) analysis is incorporated in the resulting in-context analysis, this due to the requirement that the two analyses share an identical set of operators, these being discerned in the non–contextual analysis. In consequence, the in–context (ic) analysis results in the derivation of the super–system Ss – here $Ss_{Doubling\_or\_not}$ (Figure 68). This representation has the same structure as the non-contextual (nc) analysis for *Doubling_or_not* shown in Figure 62 and provides the same capabilities, for example case based justifications. It is to be observed that both analyses necessarily have the same entry condition, namely cv. Figure 68 also shows the dynamic tracing of an application of the analysis, with one of the possible solutions of the particular problem of whether the variant **model + ing** is spelt **modelling** or **modelling**; the conditions, the algorithm branches visited, and the justifications are shown variously as **true**, **false** and *undefined* (i.e. not visited). Inspection of the conditions and operators in both analyses results (Figures 62 and Figure 68) shows that for *Doubling_or_not*, whilst the operators being the result of the non–contextual (nc) analysis concern solely morphology (as too are the conditions in the non–contextual (nc) analysis (Figures 62)), the conditions in the in–context (ic) analysis (Figure 68) range over phonetics, phonology, lexis, morphology, syntax, morpho-syntax, semantics, morpho-semantics and register (regional variation). Furthermore certain conditions can be automated simply, for example lexical conditions naming lexical items and the morphological conditions 'terminated by...'.

The index of the equivalence relation, *Eic*, determined by the algorithm, is 13.

Doubling_or_not in-context (ic) analysis - Ss$_{Doubling\_or\_not}$	
Conditions	
Id	**Condition text**
cv	**word with final consonant in English taking –ed, -ing, -er, -est, -en**
a	word of a syllable of the form C-V-C
b	**terminated by C-V-C or by C-V(pronounced)-V(pronounced)-C**
c	last syllable accented
d	**terminated by –l or –m**
e	**used in England**
f	"(un)parallel"
g	*"handicap, humbug"*
h	*"worship, kidnap"*
i	*terminated par –ic*
j	*"wool"*
Operators	
Id	Operator text
N	No doubling of the consonant
D	**Doubling of the consonant**
K	The words terminating in –ic take –ck

Algorithm represented as organigram with justifications			
Line #	Level	Condition- > Operator	Canonical -> Variant : Provenance
0	**0**	**cv -> N**	'feel->'feeling : CEOED 1971
1	1	a -> D	'run->'runner : CEOED 1971
2	**1**	**b -> N**	'answer->'answerer : CEOED 1971
3	2	c -> D	dis'til->dis'tiller : CEOED 1971
4	**2**	**d -> N**	'model->'modeling : WNCD 1981
5	**3**	**e > D**	**'model->'modelling : CEOED 1971**
6	4	f -> N	(un)'parallel-> (un)'paralleled : CEOED 1971
7	2	g -> D	'handicap->'handicapped : CEOED 1971
8	2	h -> N	'worship->'worshiped : WNCD 1981
9	3	e -> D	'worship->'worshipped : CEOED 1971
10	2	i -> K	'frolic->'frolicked : CEOED 1971
11	1	j -> N	'wool->'woolen : WNCD 1981
12	2	e -> D	'wool->'woollen : CEOED 1971

Figure 68. Representation of the in-context (ic) analysis for *Doubling_or_not* resulting in Ss$_{Doubling\_or\_not}$

Figure 69 shows the model theoretic model of the in-context (ic) analysis for *Doubling_or_not* and thus of the super-system $Ss_{Doubling\_or\_not}$ and which has the same form as that of the non-contextual (nc) analysis (see Figure 64).

0. $cv \wedge \neg a \wedge \neg b \wedge \neg j \wedge$ $N \wedge \neg D \wedge \neg K V$
1. $cv \wedge a \wedge$ $\neg N \wedge D \wedge \neg K V$
2. $cv \wedge \neg a \wedge b \wedge \neg c \wedge \neg d \wedge \neg g \wedge \neg h \wedge \neg i \wedge$ $N \wedge \neg D \wedge \neg K$
3. $cv \wedge \neg a \wedge b \wedge c \wedge$ $\neg N \wedge D \wedge \neg K V$
4. $cv \wedge \neg a \wedge b \wedge \neg c \wedge d \wedge \neg e \wedge$ $N \wedge \neg D \wedge \neg K V$
5. $cv \wedge \neg a \wedge b \wedge \neg c \wedge d \wedge e \wedge \neg f \wedge$ $\neg N \wedge D \wedge \neg K V$
6. $cv \wedge \neg a \wedge b \wedge \neg c \wedge d \wedge e \wedge f \wedge$ $N \wedge \neg D \wedge \neg K V$
7. $cv \wedge \neg a \wedge b \wedge \neg c \wedge \neg d \wedge g \wedge$ $\neg N \wedge D \wedge \neg K V$
8. $cv \wedge \neg a \wedge b \wedge \neg c \wedge \neg d \wedge \neg g \wedge h \wedge \neg e \wedge$ $N \wedge \neg D \wedge \neg K V$
9. $cv \wedge \neg a \wedge b \wedge \neg c \wedge \neg d \wedge \neg g \wedge h \wedge e \wedge$ $\neg N \wedge D \wedge \neg K V$
10. $cv \wedge \neg a \wedge b \wedge \neg c \wedge \neg d \wedge \neg g \wedge \neg h \wedge i \wedge$ $\neg N \wedge \neg D \wedge K V$
11. $cv \wedge \neg a \wedge \neg b \wedge j \wedge \neg e \wedge$ $N \wedge \neg D \wedge \neg K V$
12. $cv \wedge \neg a \wedge \neg b \wedge j \wedge e \wedge$ $\neg N \wedge D \wedge \neg K$

Figure 69. Model theoretic model of the in-context (ic) analysis for *Doubling_or_not* and thus of $Ss_{Doubling\_or\_not}$

2.2.2.2.3   *Set analysis*

Figure 70 shows an extract of the set formulation of the condition components of the in-context model theoretic model's interpretations.

Set formulation (extract) of condition components (ic)			
Algorithm		Set name	Set Value
Line #	Level		
0.	0	CV.0	$CV\backslash A\backslash B\backslash J$
5.	3	CVic.0.1.1.0	$CV\backslash A \cap B\backslash C \cap D \cap E\backslash F$
12.	2	CVic.0.2.0	$CV\backslash A\backslash B\backslash \cap J \cap E$

Figure 70. Set formulation (extract) of condition components (ic)

The sets *CVic.0*, ... *CVic.0.2.0* partition the set *CV*:

- The intersection of these sets is the empty set:
  $\cap${*CVic.0, CVic.0.0, CVic.0.1, CVic.0.1.0, CVic.0.1.1,*
  CVic.0.1.1.0, CVic.0.1.1.0.0, CVic.0.1.2, CVic.0.1.3,
  CVic.0.1.3.0, CVic.0.1.4, CVic.0.2, CVic.0.2.0 } $= \varnothing$
- The union of these sets is the set *CV*:
  $\cup${*CVic.0, CVic.0.0, CVic.0.1, CVic.0.1.0, CVic.0.1.1,*
  CVic.0.1.1.0, CVic.0.1.1.0.0, CVic.0.1.2, CVic.0.1.3,
  CVic.0.1.3.0, CVic.0.1.4, CVic.0.2, CVic.0.2.0 } $= CV$

Figure 71 shows an extract of the sets that are defined during the execution of the in–context algorithm at the point where the last condition has the value true.

Sets defined during the execution of the algorithm (ic) (extract)			
Algorithm		Set name	Set formulation
Line #	Level		
0.	0	$CV'.0$	CV
5.	3	$CV'.0.1.1.0$	$CV \backslash A \cap B \backslash C \cap D \cap E$
12.	2	$CV'.0.2.0$	$CV \backslash A \backslash B \cap J \cap E$

Figure 71. Sets defined during the execution of the algorithm (ic) (extract)

The sets thus defined form a collection of nested sets. A collection of nonempty sets is said to be *nested* if, given any pair X, Y of the sets, either $X \subseteq Y$ or $X \supseteq Y$ or X and Y are disjoint. (In other words, $X \cap Y$ is either X, Y or $\varnothing$.) (Knuth 1975: 309, 314). The non–contextual analysis of *Doubling_or_not* also results in a collection of nested sets, but there are no disjunctions. Disjunctions can occur in non–contextual analyses; for example in disambiguating applications such as se-mantic hierarchies in machine translation (Cardey et al. 2005) and ambiguous tag sets in disambiguating parts of speech taggers (Cardey & Greenfield 2003). Thus for a systemic linguistics analysis representation to be well formed, two constraints must be met: proper sub–setting and disjunction.

2.2.2.2.3.1 *Proper subsetting.* The sets defined during the execution of the in-context algorithm form a collection of proper subsets (Figure 72).

Proper subset structure (ic) (extract)	
Algorithm line #	Parent set $\supset$ Line's set
1.	$CV'.0 \supset CV'.0.0$
5.	$CV'.0.1.1 \supset CV'.0.1.1.0$
12.	$CV'.0.2 \supset CV'.0.2.0$

Figure 72. Proper subset structure (ic) (extract)

We note that in respect of nested sets this is well illustrated by the semantic feature hierarchy in the Korean to French machine translation system (Cardey et al. 2005, Cardey 2008b).

2.2.2.2.3.2 *Disjunction.* The sets defined during the execution of the in-context algorithm at the same (nesting) level and with common parent set are mutually disjoint. For example from line 0 of the algorithm:

disjoint $<CV'.0.0, CV'.0.1, CV'.0.2>$

To show this, it is necessary to show that:

$$(CV \cap A) \cap (CV \setminus A \cap B) \cap (CV \setminus A \setminus B \cap J) = \emptyset$$

It is sufficient to show that:

$$(CV \cap A) \cap (CV \setminus A \cap B) = \emptyset$$

We have:

$$(CV \cap A) \cap (CV \setminus A \cap B) = (CV \cap A) \cap ((CV \setminus A) \cap B) =$$
$$(CV \cap A) \cap (CV \setminus A) \cap B$$

But $(CV \cap A) \cap (CV \setminus A) = \emptyset$. Therefore:
$$(CV \cap A) \cap (CV \setminus A \cap B) = \emptyset \qquad \text{QED}$$

2.2.2.2.4 *In–context analysis reviewed*

To summarise in respect of the in–context analysis, this results in a partition P_{ic} over CV, thus forming an equivalence relation E_{ic} on CV, where each element of P_{ic} is an equivalence class in which each variant shares the same context dependent reasons for relating it to its corresponding canonical form ('**ic**' implying 'in-context'). The index of E_{ic}, $\#P_{ic}$, is the number of equivalence classes, this being the number of lines in the algorithm, 13 for *Doubling_or_not*. To each equivalence class the linguist attributes a unique justification, making the set J ($\#J = \#P_{ic}$). The relation which engenders this partition is $\subseteq$, the analysis leading to a nested set structure. For (disjoint) sets at the same level interval, the algorithm ensures set subtraction (by set difference) if needed (explaining the linguistically redundant ¬a, but the linguistically needed ¬c in the model theoretic model of $Ss_{Doubling\_or\_not}$ (Figure 69). For such sets at the same level interval an arbitrary (non-linguistic) relation can be added (e.g. a 'standard' ordering $\leq$).

2.2.2.2.5 *Formulation of the super-system*

We formulate the super-system Ss as a function S in the form of the binary relation between the equivalence relations E_{ic} and E_{nc}, each over CV and in particular as a surjection (see Figures 61 and 73), this latter thus ensuring both the departure set D and arrival set A (codomain or image) are totally covered respectively by the domain of definition $\mathcal{D}$ and the codomain (image) C, so providing exhaustivity.

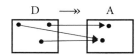

Figure 73. Surjection

Our abstract mathematical model can be resumed thus:

$$S = E_{ic} \longrightarrow\!\!\!\!\!\gg E_{nc}$$

where the infix relational operator $\longrightarrow\!\!\!\!\!\gg$ for surjection is that of the Z notation (Spivey 1992) which uses the term 'total surjection' for surjection.

We formulate the super-system in this manner for two principal reasons. Firstly, a surjection (total function onto) makes explicit the exhaustive nature of the model (both the departure set and the arrival set are covered). Secondly, being a function and in particular a surjection enables operations, whether analytic ('top–down') on a given super-system or synthetic ('bottom–up') involving super-system composition; we employ for example variously forward relational composition (see Figure 74) of super-systems and union of super-systems (as surjections) in the Labelgram disambiguating tagger system (Cardey & Greenfield 2003) which we describe in 3.2.2 Labelgram.

$$L : X \longrightarrow\!\!\!\!\!\gg Y$$
$$M : Y \longrightarrow\!\!\!\!\!\gg Z$$
$$L\,;\,M : X \longrightarrow\!\!\!\!\!\gg Z$$

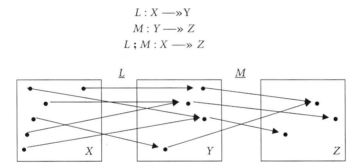

Figure 74. Forward relational composition of surjections

In Figure 75 we show $S_{Doubling\_or\_not}$, where:

$$S_{Doubling\_or\_not} = E_{Doubling\_or\_not\ ic} \longrightarrow\!\!\!\!\!\gg E_{Doubling\_or\_not\ nc}$$

in the form of the materialisation of its graph G; the justifications are also shown in each of the in-context equivalence classes in the form of corpus examples and provenances; for example ['feel > 'feeling]: CEOED 1971 being the justification for the default class, that corresponding to algorithmic line number 0.

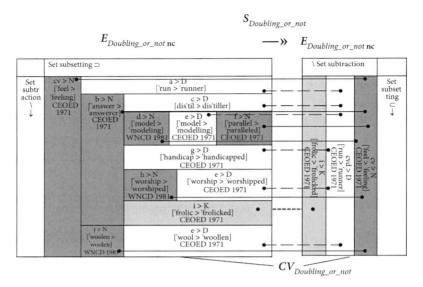

Figure 75. Graphical representation of the super-system $S_{Doubling\_or\_not}$

2.2.2.3 Optimisation considerations

The model that we have described presents optimisation possibilities in variously processing speed, space and verifiability. The nest of sets defined during the execution of the in-context algorithm at the same (nesting) level and with common parent set is formed by a process of progressive set differences; for *Doubling_or_not*, see Figure 76.

Algorithm		Set name	Set formulation
Ligne #	Level		
3	2	$CV'.0.1.0$	$CV \backslash A \cap B \cap C$
4	2	$CV'.0.1.1$	$CV \backslash A \cap B \backslash C \cap D$
7	2	$CV'.0.1.2$	$CV \backslash A \cap B \backslash C \backslash D \cap G$
8	2	$CV'.0.1.3$	$CV \backslash A \cap B \backslash C \backslash D \backslash G \cap H$
10	2	$CV'.0.1.4$	$CV \backslash A \cap B \backslash C \backslash D \backslash G \backslash H \cap I$

Figure 76. Progressive set difference process

This progressive set difference process which has as result set subtraction is safe in functional terms. We have:

disjoint $<CV'.0.1.0, CV'.0.1.1, CV'.0.1.2, CV'.0.1.3, CV'.0.1.4>$

but for a given analysis there may be variously certain redundant operations or indeed no such need of them at all. For *Doubling_or_not*, linguistic inspection of the abstraction conditions a, b, j shows that the sets of the nest A, B, J are necessarily mutually disjoint; not only is no set subtraction necessary but there is no explicit algorithmic sequencing necessary for the conditions a, b and j. However for the algorithm condition sequence <c, d, g, h, i>, whilst sets D, G, H, I are mutually disjoint, algorithmically, condition c delivering set C must for linguistic reasons precede any sequence of d, g, h, i.

There exist linguistic analysis situations where the conditions in a nesting level are such that their abstracted sets are in any case mutually disjoint; no explicit set subtraction is required. An example is the use of the form of words for raw parts of speech tagging which is explained in section '3.2.1 Morphological rule dictionary' and involves string based set subsetting leading to highly space efficient intensionally based dictionaries. The choice of nest level condition sequencing can thus depend on external criteria, such as speed optimisation in the case of automated applications. Furthermore, in such automated applications, a representation instance can be mechanically verified to be well formed in respect of proper subsetting and disjunction.

2.2.2.4 Applying the abstract mathematical model

The abstract mathematical model that we have developed can be exploited in two fashions, firstly as an aid and eventually the basis for tools for the linguist, and secondly as a means of communication between the linguist and those involved with specifying and developing applications.

It is to be understood that this abstract model is not limited and 'frozen' to the surjection:

$$S = E_{ic} \longrightarrow\!\!\!\!\!\! E_{nc}$$

which can be seen to be a finality, but to the mathematical framework and its suitable exploitation leading to this finality. To this end, mathematically, for modelling also the linguistic analysis process and the communication of this process to those involved with specifying and developing applications including tools, the starting point is the empty binary relation leading to the the model of the classification of binary relations, for which a particular algorithmic instance has been shown in Figure 61.

We observe that both the model theoretical models (non-contextual (nc) and in−context (ic)) have at most one negation apposed to a condition, such negations being implicit in the organigram representation when passing to a subsequent condition at the same nesting level. In applying our methodology, at the algorithmic representation level we avoid conditions which are explicit negations involving open sets. Doing so avoids disallowing a constructive modelling approach as it ensures that there are no resultant double negations in the model theoretic models. Such constructive modelling leads to the possibility of employing formal methods whose use can be obligatory for computational applications in certain safety critical domains.

As a means of communication between the linguist and those involved with specifying and developing applications, the abstract mathematical model acts as a fulcrum; for example the synthesis of micro-systemic linguistic analysis and software engineering is illustrated in Figure 77 (Cardey & Greenfield 2008).

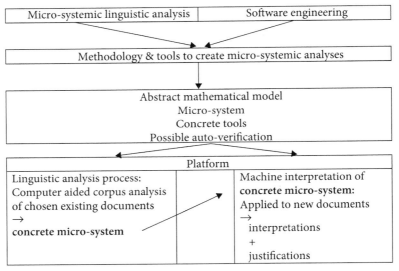

Figure 77. The pivotal nature of the abstract mathematical model as fulcrum between the linguist and the software engineer

2.2.2.4.1   *Model–driven evaluation*

Concerning application evaluation, our theory enables a model–driven evaluation procedure which was initially developed and itself evaluated in an agro-food industry application (Cardey et al. 2006). This procedure has the benefit of incorporating case-based testing and the production of a complementary exhaustive benchmark (see Figure 78).

When we say 'model–driven procedure' we mean that this procedure itself is a concrete realisation of our abstract mathematical model but here applied to the domain of systemic linguistic analysis itself, this in respect of the evaluation of a particular systemic linguistic analysis application. For example in Figure 78 the column pair:

Model–driven evaluation procedure				
Recognition	Nature of context		Demonstrates	Action
Correctly recognised	Corpus attested context		Specific analysis	None – success
	Linguist's competence context	Attested category in other context(s) (e.g. in same corpus)	Generality	Add attestation to context & to automatic case-based benchmark datum
		Category not attested in e.g. same corpus, but attested in other corpus/a	Cross corpus generality	Add attestation to context & to automatic case-based benchmark datum
Error – Not recognised	Lack of cover: category and/or context missing		Location of error	Insert category and/or context, do regression test
Error – Incorrectly recognised	Context error		Location of error	Correction of context, do regression test

Figure 78. Model–driven evaluation procedure

| Recognition | Nature of context |

corresponds to the evaluation algorithm with its conditions and the column pair:

| Demonstrates | Action |

to the evaluation operations. In the evaluation procedure, the cases correspond one–to–one to variously the application model's in-context equivalence classes (the 'categories' in Figure 78).

To carry out the evaluation of some application that we are developing, in following the evaluation procedure we initially construct the raw input working corpus, partitioning this into the initial data set enabling 'boot–strapping' of the analysis, and sample test data sets. These latter are subsequently used in the incremental regression testing, instrumentation and manual (i.e. the linguist) evaluation sub–tasks (involving for example stability/asymptotic criteria satisfaction).

PART 3

Methodologies and applications

The core model has given rise to methodologies and applications (Cardey & Greenfield 2006) such as disambiguated parts of speech tagging – the Labelgram system (Cardey & Greenfield 2003), machine translation of 'far' language pairs (including anaphoric reference and zero anaphora processing) (Cardey, Greenfeld & Hong 2003; Alsharaf et al. 2004a, Alsharaf et al. 2004b), grammar checking and correcting (including noun phrase identification) (Vienney et al. 2004), 'sense' mining or concept mining (seme mining) (Cardey et al. 2006) 'The Classificatim Sense-Mining System', and for safety critical applications where evaluation ability is required in the form of validation and traceability as in controlled languages, for example cockpit alarm message vocabulary (Spaggiari et al. 2005), in the machine translation of medical protocols (Cardey, Greenfield & Wu 2004), and also in automatic acronym recognition in safety critical technical documentation (Cardey et al. 2009).

We will show how, in applying the theory "micro-systemic analysis", we can solve difficult problems.

Grammar checkers

We start by returning to our example of the application of micro-systemic theory to French lexico-morpho-syntax for the agreement of the French past participle. We have seen that syntax cannot be in reality separated from the lexis or from morphology. We have shown how such a system can be represented so that it can be processed by machine and how another methodology derived from the same theory is used for solving the problem of how to do the agreement in function with the form of the lexical items for the past participle followed by an infinitive.

We will take therefore yet another sub-system of our system for the agreement of the French past participle, this being the agreement of the past participle of pronominal verbs.

We begin with a corpus composed of a representative of each class of the partition of the pronominal verbs due to Yves Gentilhomme which we reproduce here.

> ### Les amours de Jeanneton et Jeannot
>
> Jeanneton s'est blessé()1, elle s'est entaillé()2 profondément deux phalangettes avec une faucille. C'est à sa main gauche qu'elle s'est coupé()3 pendant qu'elle s'était penché()4 sur un tas de joncs. S'étant vu()5 saigner abondamment – sa robe, d'ailleurs, s'en était trouvé()6 toute tâché()7 et fripé()8, elle s'était cru()9 mourir, mais elle s'est ressaisi()10 et s'est comporté()11 courageusement. Elle s'est souvenu()12 de son ami Jeannot, s'est mis()13 à l'appeler, puis s'est laissé()14 transporter par lui à la clinique.
>
> Elle s'en est tiré()15 à bon compte, s'était-il entendu()16 dire par les infirmiers auprès desquels elle s'est fait()17 soigner.
>
> Par la suite les mauvaises langues se sont délié()18. Bien des histoires se sont raconté()19. Il s'en serait passé()20 des choses! Nous, ses copines, on s'est cru()21 en plein délire et moi, je ne m'en suis pas encore remi()22, à parler franc.
>
> De bon matin, Jeanneton et Jeannot se seraient levé()23, puis après s'être trop coquettement habillé()24, lavé()25, coiffé()26 et enfin, chacun de son côté, dirigé()27 vers le marais, où la veille ils se seraient entendu()28 pour se revoir, ils s'y seraient secrètement retrouvé()29. Par de vilaines mégères, ne s'était-il pas dit()30 que ces blancs-becs se seraient ri()31 des conseils qu'ils s'étaient entendu()32 prodiguer ? Hélas, ils se seraient complu()33 dans leur entêtement, ils se seraient amusé()34, se seraient joué()35 de tous et se seraient plu()36 à contredire leurs parents.
>
> Comment ces commères s'y sont-elles pris()37 pour s'en être persuadé()38 jusqu'à s'être adonné()39 à de telles calomnies ?

D'ailleurs personne, pas même la femme du garde champêtre, ne s'est aperçu()40 de rien, laquelle pourtant ne s'est jamais ni privé()41 de médire, ni gêné()42 de faire courir des ragots.

D'aucuns se seraient fait()43 confirmer qu'après s'être bien conté()44 fleurettes, les deux se seraient fait()45 pincer s'ils ne s'étaient pas douté()46 du danger. Bref ils se sont méfié()47 et se sont esquivé()48 à temps. Aussi personne ne les a vus ni s'être dissipé()49, encore moins s'être enlacé()50 ou s'être abandonné()51 à des actes coupables.

Après ces émotions l'un comme l'autre se sont retrouvé()52 assez fatigué()53, s'étant beaucoup dépensé()54 et même un peu énervé()55 sans cependant s'être querellé()56 ni grossièrement insulté()57.

L'aventure finie, des retombées, ne s'en serait-il ensuivi()58 ?

Bien que la morale de l'histoire ne se soit prononcé()59 que sur le fait que les joncs se sont raréfié()60, je me suis cru()61 obligé()62 de conclure que les deux jouvenceaux se sont dûment marié()63 et même se sont perpétué()64 par de nombreux héritiers. Ah, que d'émotions !

Son amie Larirette

The above text represents all the possibilities for the agreement of the past participle of French pronominal verbs.

How do we go about doing the agreements for this text?

The methodology for writing our algorithm can be organised in the same manner as in the example already given concerning the agreement of the French past participle followed by an infinitive.

Let us take the first sentence of our text to see if we can check its past participles automatically:

(1)   *Jeanneton s'est blessé()1, elle s'est entaillé()2 profondément deux phalangettes avec une faucille.*

(Jeanneton has injured herself, she has deeply gashed her distal phalanges/ finger tips with a sickle.)

The first step of our grammar checker is the tagging of each unit of the sentence; this is performed by a morpho-syntactic analysis. We use the automatic morphological disambiguated tagging of texts system, Labelgram (Cardey et al. 1997), (Cardey & Greenfield 1997), (Cardey & Greenfield 2003), which has been developed in the L. Tesnière Research Centre. This system is also based on a methodology developed from our same theory. The strength of this system is its intensionally driven approach leading not only to representations which are efficient in terms of size (Cardey et al. 1997) but which accept for example, in the case of word dictionaries, neologisms 'obeying the rules for French word formation'.

Moreover, Labelgram disambiguates each unit of the sentence; a first super-system finds all the grammatical categories (parts of speech) of each unit and a second super-system calculates the right category in the given text, using context rules. This second super-system is related to the first super-system by means of a model of the class ambiguities in French in respect of parts of speech.

For example, the unit "*blessé*" in our sentence can be:

- a Substantive: *Le **blessé** est conduit à l'hôpital.* (The wounded is led to the hospital.)
- a Past participle: *Le chasseur a **blessé** l'animal.* (The hunter wounded the animal.)
- an Adjective: *L'homme **blessé** et fatigué ne peut pas courir.* (The wounded and tired man cannot run.)

In Figure 79 we show the result of Labelgram's analysis of the first sentence of *Les amours de Jeanneton et Jeannot*:

> *Jeanneton s'est blessé()1, elle s'est entaillé()2 profondément deux phalangettes avec une faucille.*

In order to simplify the syntactic analysis for the automatic checking of the agreement of the past participle, we only keep the elements of the sentence necessary for our application; in this way we reduce the sentence to its 'minimum'. This means

Mot	Catégories	Catégorie
Jeanneton	[Nom]	Nom
s	[Pro. pers.]	Pro. pers.
est	[Nom,Verbe conj.]	Verbe conj.
blessé	[Nom,Ppa.,Adj. verbe]	Ppa.
,	[virgule]	virgule
elle	[Pro. pers.]	Pro. pers.
s	[Pro. pers.]	Pro. pers.
est	[Nom,Verbe conj.]	Verbe conj.
entaillé	[Ppa.,Adj. verbal]	Ppa.
profondément	[Adv.]	Adv.
deux	[Art.]	Art.
phalangettes	[Nom]	Nom
avec	[Adv., Prép.]	Prép.
une	[Art. ind.,Adj. num. card.,Adj. num. ord.,Nom]	Art.ind.
faucille	[Nom]	Nom
-	[point]	point

Figure 79. Result of the Labelgram analysis of the first sentence of *Les amours de Jeanneton et Jeannot*

that we keep the sentence's subject and verb, and sometimes its objects. For this particular application, we just have to remove all what appears between parentheses in our algorithm.

If we take the complete structure:

Pl(,)(Adv)P2(Adv)(I')(P3)(Adv)(P4)(Adv)avoir(Adj)(P3)(P4)(Adv)
List(Adv)(Prép)(P5)

we keep:

PlP2avoir**List**

For our example sentence, the system detects two past participles (Ppa.) (see the Result of the Labelgram analysis in Figure 79) to check: *blessé* et *entaillé*. It should thus find two structures.

The 'superficial' units of the sentence correspond to the elements in parentheses in our system's algorithm: the adjectives, the adverbs, but also the incidental complements, etc. In reality, all these elements have no influence on the agreement of the past participle. Their role is to specify variously the place, the manner and the time of the action. These are precisions for which our current application has no interest. This is the reason why they can be separated out and removed by our grammar correcting system.

Thus, for our initial sentence:

(1)   *Jeanneton s'est blessé()1, elle s'est entaillé()2* **profondément** *deux phalangettes avec une faucille.*

we obtain after the simplified system, the following two 'sub-sentences':

(1)   a.   *Jeanneton s'est blessé*
       b.   *elle s'est entaillé deux phalangettes*

These two sub-sentences correspond to the following two syntactic structures respectively:

(1)   a.   *P1 (Pro.pers.) (Verbe conj.) (Ppa.)*
       b.   *P1 (Pro. Pers.) (Verbe conj.) (Ppa.)* **P2**

Then, our system applies the agreement rule linked to each of the above structures in order to check the past participle and correct it if needs be.

For the first structure:

(1)   a.   *Jeanneton s'est blessé*
       a.   *P1 (Pro.pers.) (Verbe conj.) (Ppa.)*

the system contains the rule:

P1 Pro.pers. être Ppa. **Agreement with P1**

Thus, the system checks that the Ppa. *blessé* is in correct agreement in gender and number with the subject P1 *Jeanneton*:

| *Jeanneton* | gender: | **feminine** | number: | **singular** |
| *blessé* | gender: | <u>**masculine**</u> | number: | **singular** |

The system detects an error and corrects it automatically:

(1) a. *Jeanneton s'est blessé<u>e</u>*

For the second structure:

(1) b. *elle s'est entaillé deux phalangettes*
   b. *P1 (Pro. Pers.) (Verbe conj.) (Ppa.) P2*

our system contains the rule:

P1 Pro.pers. être Ppa. P2 **Invariable**

Thus, the system checks that the Ppa. *entaillé* is in the invariable form (operator **I**), which means masculine, singular:

| *entaillé* | gender: | **masculine** | number: | **singular** |

The past participle does not need any correction. The system continues with the next sentence.

As we can see for this application the first step was to determine the grammatical category of each of the sentence's lexical units.

Part of speech tagger

3.2.1 Morphological rule dictionary

We are now going to see how our methodology which enables the application of determining grammatical categories is founded on the same theoretical model.

This particular application provides an example of what it means to work in intension as opposed to in extension.

As we have already explained in Part 1, it is necessary at the outset to determine the parts of speech that one needs knowing that furthermore these vary from one language to another.

Each of the lexical items of the language to be processed ought then to be linked to one (or more in case of ambiguity) parts of speech and be entered in a dictionary.

Having defined our parts of speech, rather than placing, as we find in electronic dictionaries, whatever their function, as headwords (entry words) each of the forms, canonical and inflected associated with one or more part of speech, our dictionary presents the headwords in the form of morphological characteristics classified in sets and subsets. These characteristics are thus classified according to certain constraints and are associated with one or more of the parts of speech that we need for the required applications.

What characteristics are to be chosen?

For a long time, the majority of grammar manuals and linguistic treatises on French for example have mentioned that certain suffixes give an indication as to the category to which a 'word' belongs. Certain suffixes serve to form rather nouns, others rather adverbs etc.

For example, for French:

- Suffixes serving to form nouns
    - -ace   populace
    - -ade   orangeade
    - -age   balayage
- Suffixes serving to form adjectives
    - -able, -ible   aimable, audible
    - -aire   solaire, polaire
    - -ique   chimique, ironique

- Suffix serving to form adverbs
  -ment   gentiment
- Suffixes serving to form verbs
  -ailler   tournailler
  -asser   rêvasser

Suffixes are thus often classified according to the nature of the words that they serve to form. We have been tempted to experiment with a classification not based uniquely on suffixes in the traditional sense but on what we call 'endings'. Each ending determines to what grammatical category(ies) or part(s) of speech (these two designating the same concept) the lexical items belong.

Thus our dictionary enables recognising the grammatical category(ies) of all 'words' by analysing their endings.

For the languages which have interested us (for example English, French, German and Spanish), we have thus partitioned the lexis into sets as much as possible. If a sequence of graphemes corresponded to a grammatical category/part of speech for a sufficiently large part of the lexis (canonical and inflected forms), we kept it. However we have had to keep the exceptions and develop what we call rules and sub-rules.

For example, for French, take the ending -aire.

Ignoring the exceptions we have:

-aire	→	noun
-baire	→	adjective
-faire	→	verb in the infinitive
-iaire	→	adjective
-laire	→	adjective

Now take the lexical items ending in -able.

Our general rule says that the words ending in -able are adjectives:

-able	→	adjective

We have the following lists of exceptions:

{cartable,...}	→	noun
{accable,...}	→	conjugated verb
{sable, table,...}	→	noun or conjugated verb
{comptable,...}	→	noun or adjective

We see appearing here the problem of ambiguities to which we will subsequently return. Lexical items like *table, comptable* can belong to two parts of speech.

Words ending in–ague   → noun

    List of exceptions
    {bague, blague, dague, drague}   → noun, conjugated verb

    List of exceptions
    {vague}   → noun, adjective, conjugated verb

    List of exceptions
    {rague, zigzague}   → conjugated verb

Figure 80. The ending -ague only to be found in our dictionary

Our endings system has little to do with the derivational system to which we are accustomed. The example in Figure 80 shows an ending that is only to be found in our dictionary.

From these examples we see that the dictionary is constituted with rules. A general rule can be accompanied by sub-rules. A sub-rule can be composed with an ending depending on the principal ending as -baire which depends on -aire. A sub-rule can also be simply composed of a list of 'words' ending with the same graphemes exactly as those of the principal ending. Each sub-rule is therefore composed of an ending or a list associated with one or more grammatical categories.

If there is more than one grammatical category, we have an ambiguity. As in our algorithms, the sub-rules obey a particular reading and thus processing order. The general rules can be read in any order.

We take an extract from the algorithm for the French parts of speech tagger which has been formulated, as has already been explained, in sets and partitions. See Figure 81, where 'Mot se terminant par' = 'Word ending in', 'Consulter la liste' = 'List of exceptions', and the names of the parts of speech are in French.

Take the word *cheval* which we are going to process with our algorithm.

We have a general rule (21) which says that all lexical items ending in -al belong to the category Adj. (adjective).

However, before applying this rule, we have to consult what are called the sub-rules.

We thus follow the following path:

    21 true, 21.1 false, 21.2 false, 21.3 false and 21.4 true;
    conclusion *cheval* is a nom (noun)

For *frugal*:
    we have:

    21 true, 21.1 false, 21.2 false, 21.3 false, 21.4 false, 21.5 false, 21.6 false; therefore the last true condition being 21, we apply the operator Adj (adjective).

21/ Mot se terminant par al	Adj.
21.1/ Mot se terminant par aal	Nom
21.2/ Mot se terminant par hal	Nom
21.2.1/ Consulter la liste	Adj.
catarrhal; nymphal; triomphal; zénithal	
21.3/ Mot se terminant par oal	Nom
21.4/ Consulter la liste	Nom
amiral; ammonal; arsenal; aval; bacchanal; bal; bocal; cal; canal; caporal; caracal; carnaval; cérémonial; chacal; chenal; cheval; chloral; confessionnal; copal; corporal; dispersal; diurnal; fanal; fécial; festival; foiral; gal; galgal; gavial; géosynclinal; grémial; journal; madrigal; majoral; mémorial; mistral; monial; narval; nopal; official; orignal; pal; pipéronal; prairial; processionnal; racinal; rational; régal; revival; rorqual; salicional; sandal; serval; sial; signal; sisal; tergal; tincal; tribunal; urinal; val	
21.5/ Consulter la liste	Nom, Adj.
animal; anormal; anticlérical; anticlinal; armorial; bancal; brachial; brutal; capital; cardinal; central; clérical; colonial; commensal; cordial; diagonal; dorsal; éditorial; égal; épiscopal; équatorial; final; fromental; général; germinal; international; libéral; littoral; local; méridional; minéral; moral; oral; original; pascal; pectoral; pontifical; présidial; principal; provincial; radical; réal; régional; rival; sidéral; social; spiral; synclinal; vassal; vertical; vespéral	
21.6/ Consulter la liste	Nom, Adj., Adv.
mal	

Figure 81. Extract from the algorithm for the French parts of speech tagger

For *triumphal*:
   we have:

   21 true, 21.1 faux, 21.2 true, 21.2.1 true;
   therefore *triomphal* is an Adj (adjective)

This formalisation avoids any confusion between the endings. This hierarchisation within a rule is valid for humans as well as machines.

What does this form of dictionary contribute?

Let us see the advantages.

We have a set of 579 endings accompanied by sub-endings (-aire, -baire for example) or lists of exceptions.

Only the lists of exceptions appear in detail and this is a gain in processing and in size. But we are going to see even greater advantages in what follows.

We take as example rule 160 which recognises the forms in -er; see Figure 82.

Ending-er and its exceptions				Operator	Automatic Dictionary	Le Robert
er				*Verbe inf.*	1	4932
	ier			*Nom*	1	1025
		fier		*Verbe inf.*	1	125
			Liste	*Nom*	3	3
			Liste	*Adj.*	2	2
			Liste	*Nom, Adj.*	1	1
		gier		*Verbe inf.*	1	5
			Liste	*Nom, Adj.*	1	1
	Liste			*Adj.*	42	42
	Liste			*Nom, Adj.*	144	144
	Liste			*Verbe inf.*	127	127
	Liste			*Nom, Verbe inf.*	5	5
	Liste			*Adv.*	3	3
Liste				*Nom*	184	184
Liste				*Adj.*	9	9
Liste				*Nom, Verbe inf.*	41	41
Liste				*Nom, Adj.*	13	13
Liste				*Verbe inf., Adj.*	1	1
Liste				*Adv.*	1	1
Liste				*Nom, Adj., Adv.*	1	1
Liste				*Nom, Adj., Verbe inf.*	1	1
Total					583	6666

Figure 82. Comparison in terms of number of entries between our dictionary and a conventional dictionary

In Figure 82, the figures are the number of lexical items. A figure near to zero in our dictionary indicates that the entry is well chosen. Inversely, in conventional, that is, extensional dictionaries, the larger the figure then the greater the processing.

We have taken the Le Robert electronic dictionary of French (1997) and have counted 6666 entries for the ending -er and only 583 for our dictionary.

Another advantage and not the least is that our dictionary allows listing the French morphological ambiguities.

We observe in the algorithmic micro-system shown in Figure 83 that in certain of the sub-rules we have several grammatical categories.

In Figure 83, sub-rules 6.1, 6.2 and 6.4 concern all the words in -ace which are polycategorial.

Our rules based dictionary enables recognising non ambiguous 'words' and polycatgorial 'words'. Our dictionary has brought to the foreground 29 types of morphological ambiguity amongst which certain are shown in Figure 84.

6/ Mot se terminant par **ace**   **Nom**
    6.1/ Consulter la liste   **Nom, Verbe conj.**
        agace; dédicace; espace; glace; grimace; menace;
        place; préface; surface; trace
    6.2/ Consulter la liste   **Nom, Adj.**
        biplace; boniface; contumace; dace; monoplace;
        rapace
    6.3/ Consulter la liste   **Adj.**
        coriace; efficace; fugace; inefficace; perspicace;
        sagace; salace; tenace; vorace
    6.4/ Consulter la liste   **Adj., Adv.**
        vivace
    6.5/ Consulter la liste   **Verbe conj.**
        déglace; délace; déplace; désenlace; efface; enlace;
        entrelace; lace; remplace; replace; retrace;
        surglace; verglace

Figure 83. Our dictionary: algorithmic micro-system showing certain sub-rules with several grammatical categories

Nom - Participe passé - Verbe conjugué   *(reçus)*
Nom - Adverbe -Verbe conjugué   *(soit)*
Nom - Adverbe - Adjectif - Verbe conjugué   *(ferme, double, trouble)*
Nom - Adjectif - Verbe conjugué - Interjection   *(fixe)*

Figure 84. Our dictionary: examples of morphological ambiguities

la petite joue au ballon
(the little girl plays with the ball)

la   sous-règle 2.12 *(Art., Nom, Pro. Pers)*
petite   sous-règle 281.4 *(Nom, Adj.)*
joue   sous-règle 398.1 *(Nom, Verbe conj.)*
au   sous-règle 475.1.5 *(Art.)*
ballon   règle 368 *(Nom)*

Figure 85. Example of sentence that has been analysed by our French morphological dictionary

To conlude concerning our French morphological dictionary, in Figure 85 we show a sentence that has been analysed by the dictionary.

This same methodology has been used for determining the gender of French nouns. This has allowed us to obtain a complete system which will give for example for French:

- nouns ending with the suffixes -ier, -age, -illon are masculine
- nouns ending with the suffixes -ade, -aie, -ée are feminine (with sub-rules of exceptions)

This methodology, based on our theory, has been used as the starting point for creating a system for raising morphological ambiguities.

Our French morphological dictionary enables tagging the following French sentence in respect of parts of speech:

la méchante rigole car le petit est malade
(the nasty woman laughs because the little girl is ill)

as is shown in Figure 86 in the column 'Categories'.

In Figure 86 we highlight (in bold) the first step of a system for disambiguating parts of speech in French sentences. This first step involves our French morphological dictionary system which tags giving all the parts of speech possible (columns **Categories** and **Dict. ref**). Another system completes our rule-based morphological dictionary by performing the disambiguation. This latter system too is a rule-based dictionary.

3.2.2 Labelgram

Labelgram is a system for the automatic morphological disambiguated tagging of texts (Cardey et al. 1997, Cardey & Greenfield 1997, Cardey & Greenfield 2003).

In Labelgram, the overall systemic grammars are 'super' micro-systems modelling the relationship between texts and morphologically disambiguated tagged texts. The source and variant system is the language (in the form of texts), the target and invariant system is the morphological disambiguated tagging of each text unit, and the relationship between the two is the (composite) system of linguistic data which implements this relationship.

Word form	Categories	Category	Dict. ref.	Proc n^0	Rule ref.
la	[Art.,Nom,Pro. pers.]	Art.	2.12/	5	45/
méchante	[Nom,Adj.]	Nom	41.2/	28	339/
rigole	[Nom,Verbe conj.]	Verbe	360.4/	8	79/
car	[Nom,Conj.]	Conj.	47.5/	10	144/
le	[Art.,Pro. pers.]	Art.	pre_dict	5	45/
petit	[Nom,Adj.]	Nom	279.1/	28	339/
est	[Nom,Verbe conj.]	Verbe	pre_dict	8	74/
malade	[Nom,Adj.]	Adj.	13.1/	28	346a/

Figure 86. The first step of a system for disambiguating parts of speech in French sentences

In the representation of a base micro-system relationship, a context rule consists of a description of linguistic units which is so organised so as to have a specific linguistic categorisation (of the type desired by the linguist). The nested form of context rules, each being either a set comprehension (in intension) for regularities (defaults) or an enumeration (in extension) for exceptions, results in an intensionally driven approach leading not only to representations which are efficient in terms of size (Cardey et al. 1997) but which accept for example, in the case of word dictionaries, neologisms 'obeying the rules for word formation'.

In Labelgram, the text under analysis is in the form of sequences of 'units' where a unit, depending on the analysis, may be automatically annotated (e.g. tagged with grammatical category(ies)). For Labelgram, units include:

- individual letters where the 'text' is a 'word'; used for morphological phenomena in isolation, e.g. finding grammatical category(ies) as a function of the word's ending, as in the -er example (Figure 82).
- individual 'word' forms or compounds (treated as single entities); e.g. for grammatical category disambiguation as a function of tagged word forms/compounds in context with neighbouring units. Punctuation, text and sentence boundaries are also so classified.
- 'compounds'; a compound is treated as a single entity, to be tagged (before any disambiguation) with one or more grammatical categories. For example, for French, in:

> Il est arrivé *juste à temps*
> (he arrived just in time)

the machine can calculate that *juste à temps* is a compound, in fact an adverbial locution, that is with the function of an adverb. Compounds can themselves also be polycategorial. English compounds of the type preposition + determiner + noun are usually either adjectives or adverbs. Consider:

> on the nose

which can be either an adjective ('annoying') or an adverb ('precisely').

3.2.3   Applications to different languages

We have applied the Labelgram methodology to different languages.

3.2.3.1   German

We give an example for German due to Anne-Laure Venet for the morphological ambiguity 'Appenzeller':

(1) Der *Appenzeller Bezirk* entwickelt seine Industrie.
(The Appenzell region develops its industry.)

(2) Die *Appenzeller* haben den Ruf entgegenkommend sein.
(The people of Appenzell have the reputation of being welcoming.)

In (1) *Appenzeller* is an adjective and in (2) a substantive. Raw tagging is based on a micro-system comprising a combination of closed sets (determiners, pronouns, conjunctions, interjections, etc.) and the same technique for French based on word form endings. *Bezirk* is classified as unicategorial {Subst.} and *Appenzeller* polycategorial {Adj., Subst.}. For polycategorial word forms recourse is made to another micro-system in which contexts are over sequences of elements in the sentence. For resolving the ambiguity on *Appenzeller* refer to the algorithmic micro-system in Figure 87, in which 1.1 and 1.2 are order dependent. AF indicates the ambiguous word form (here *Appenzeller*), -ler indicates a word form ending in *ler* and ^– indicates any word form starting with a capital letter.

In German a substantive never follows a substantive (except if separated by a punctuation sign which is not so here). A word starting with a capital letter followed by a substantive is an adjective. For (1), *Appenzeller* will be tagged by the algorithmic micro-system (Figure 87) as an adjective by context rule 1.2 (having passed 1., and 1.1 having failed), and for (2) as a substantive by the 'default' context rule 1., neither sub context rule 1.1 nor 1.2 being applicable.

As will be seen in the algorithmic micro-system (Figure 87), context rules themselves operate in context with other context rules, their contexts as conditions and thus the context rules themselves are nested, and there are no explicit logical negations neither within a context (the elements of which being themselves micro-systems for example SUBST) nor apposed to a context. This practice is in line with what we have said in applying the abstract mathematical model so as to ensure constructive modelling.

The appearance of the context rules as a whole resembles a concordance, not of text but of grammatical contexts in the micro-system's Source (Departure) where each item being concorded corresponds to a category in the micro-system's Target (Arrival), the item's structural position in the grammatical context being

> 1. AF(-ler) –> Subst. : 'Die *Appenzeller* haben den Ruf
>   entgegenkommend sein.'
>> 1.1 set of exceptions in the form of (AF(WORD_FORM)–>
>>   CATEGORY : CORPUS_EG)
>> 1.2 AF(^–) + SUBST –> Adj. : 'Der *Appenzeller* Bezirk
>>   entwickelt seine Industrie.'

Figure 87. Algorithmic micro-system for resolving the ambiguity on *Appenzeller*

indicated by a marker whose name itself indicates the linguistic phenomenon being concorded (here 'AF'), and all this in the 'context' of marked (e.g. *Appenzeller*) corpus examples. Compile time verification of well formedness can be effected to ensure that for example in the set of exceptions 1.1 each and all the word forms do end in *ler* (proper subset of 1.). Also, 1.1 can be optimised so that in their compiled form, the word forms have their *ler* endings removed (redundant constraint having passed the default context constraint in 1.).

3.2.3.2   Spanish

In the following Spanish example due to Helena Morgadinho:

> El **almuerzo** en **este** restaurante fue delicioso.
> (The lunch in this restaurant was delicious.)

out of context, *almuerzo* and *este* are polycategorial. Raw tagging (similar to French and German) produces:

Word form	El	almuerzo	en	este	restaurante	fue	delicioso
Categories	{Art.}	{Sust., Vbo. conj.}	{Prep.}	{Sust., Adj., Adj. demons.}	{Sust.}	{Vbo. conj.}	{Adj.}

Disambiguation starts with {Sust., Adj., Adj. demons.} (*este*). Knowing that although *en* can be followed by several grammatical categories, it is also the case that concerning this type of ambiguity:

- *en* cannot be followed by an Adj., thus one category less;
- *en* can be followed by a Sust., but this is not possible with another Sust. following (*restaurante*), (at least a separating comma would be needed which is not the case here).

From this, the following context rule can be formulated, where AF indicates the ambiguous form (here *este*):

> {en} + AF + SUST → Adj. demons.: 'El almuerzo en *este* restaurante fue delicioso.'

Disambiguation is completed by resolving {Sust., Vbo. conj.} (*almuerzo*). Now, various grammatical categories can be found before a Vbo. conj., even punctuation signs, but never articles, so this leaves just the one possibility for *almuerzo*, Sust, giving the following context rule:

> ART + AF → Sust.: 'El *almuerzo* en este restaurante fue delicioso.'

The disambiguated tagging is thus:

Word form	El	almuerzo	en	este	restaurante	fue	delicioso
Category	{Art.}	{Sust.}	{Prep.}	{Adj demons.}	{Sust.}	{Vbo conj.}	{Adj.}

3.2.3.3   English

In the following example due to Gaëlle Birocheau (2000):

The man who works here lives *in* London.

with raw tagging similar to the previous languages, followed by disambiguating context rules gives:

Word form	The	man	who	works	here	lives	in	London	
Raw categories	{DETsing, DETplu}	{N, V}	{PROinterr, PROrel}	{Nplu, V3}	{ADV}	{Nplu, V3}	{PREP, ADV, ADJ, N}	{N}	{PUNCT}
Results of Disambiguation	{DETsing}	{N}	{PROrel}	{V3}	{ADV}	{V3}	{PREP, ADV, ADJ, N}	{N}	{PUNCT}

For *in*, in the absence of context rules dealing with composition or verbal complements, the four combinations

$$V + \mathbf{PREP} + N, \quad V + \mathbf{ADV} + N, \quad V + \mathbf{ADJ} + N, \quad V + \mathbf{N} + N,$$

are all valid.

3.2.3.4   French

We have seen in the preceding paragraphs raw tagging using the morphological dictionary.

In the following sentence due to Zahra El Harouchy all the words are ambiguous.

la méchante rigole car le petit est malade
(the nasty woman laughs because the little boy is ill)

All the word forms are polycategorial and the raising of the ambiguities is initially due to resolving {Nom,Conj.} on the word *car*. For this sentence as input, Labelgram gives the output shown in Figure 88.

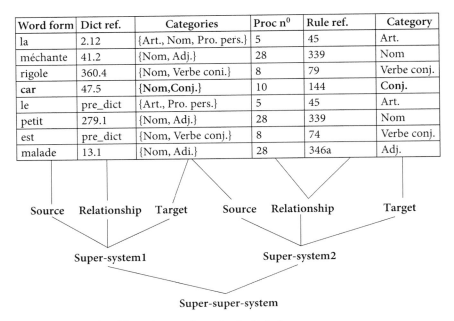

Word form	Dict ref.	Categories	Proc n⁰	Rule ref.	Category
la	2.12	{Art., Nom, Pro. pers.}	5	45	Art.
méchante	41.2	{Nom, Adj.}	28	339	Nom
rigole	360.4	{Nom, Verbe coni.}	8	79	Verbe conj.
car	47.5	{Nom,Conj.}	10	144	Conj.
le	pre_dict	{Art., Pro. pers.}	5	45	Art.
petit	279.1	{Nom, Adj.}	28	339	Nom
est	pre_dict	{Nom, Verbe conj.}	8	74	Verbe conj.
malade	13.1	{Nom, Adi.}	28	346a	Adj.

Source   Relationship   Target   Source   Relationship   Target

Super-system1   Super-system2

Super-super-system

Figure 88. Example of disambiguated tagging by Labelgram

The two principal micro-systems constituting Labelgram are:

- Super-system1: raw tagger dictionary algorithmic micro-system providing the non ambiguous and ambiguous terms of French,
- Super-system2: the disambiguator comprising the part of speech ambiguities of French; that is the class ambiguities which are linked to the set of possible French syntactic structures in the form of algorithmic micro-systems.

We can model Labelgram thus:
We declare:

S_LU to be the set of sentences segmented in lexical units;
S_LU_RT to be the set of sentences segmented in lexical units with each unit raw tagged;
S_LU_DT be the set of sentences segmented in lexical units with each unit tagged with a single category.

We now declare as surjections:

$Super\_System1$: $S\_LU \longrightarrow S\_RT$ Raw tagger
$Super\_System2$: $S\_RT \longrightarrow S\_DT$ Disambiguator
$Super\_Super\_System$: $S\_LU \longrightarrow S\_LU\_DT$ Labelgram

Using forward relational composition, we construct an instance of *Super_Super_System*, that is, Labelgram:

$$Super\_Super\_System = Super\_System1\ ;\ Super\_System2$$

which keeps the properties of its two components of being a surjection.

The raw tagger and the disambiguator are formed themselves of micro-systems which are identified by reference numbers (this can be seen in Figure 88).

The Labelgram system gives the trace of the disambiguation. If one takes the word *car* (meaning *because* or *bus*) the raw tagger micro-system gives a tag with two possible parts of speech out of context due to the rule number 47.5 and the disambiguating micro-system disambiguates in context by means of its ambiguity class micro-system 10 and sub-micro-system 144.

We can see the identification of the specific rules used by the two principal micro-systems for solving the disambiguisation. This is another advantage of the methodology which allows tracing of the operations that have been applied. In the case of an error, correcting the error is simplified because one knows the location of the problem causing the error, and thus where to effect the problem's solution. Our knowledge that the raising of the ambiguities is initially due to resolving {Nom,Conj.} on the word *car* is due to Labelgram's tracing.

A still further advantage of the methodology is that for automatic applications like Labelgram which depend solely on form, the well-formedness of the algorithmic micro-systems can be mechanically verified.

Finally, Labelgram has the possibility of tagging syntactically ambiguous sentences such as:

> la petite brise la glace.
> (the little girl breaks the ice,
> the light breeze freezes her)

3.2.4 Benchmarking

A particular bugbear in human language technology applications concerns the provision of static traceability and thus comprehensive benchmarks which are of adequate quality. Because a micro-systemic linguistic analysis is case based, it should be comprehensive (correct and complete), and thus for it the word 'adequate' has a clear meaning. The point here is that because Labelgram is based on micro-systemic linguistic analysis, a benchmark corpus is intrinsic to Labelgram in its linguistically developed and validated form. Now, Labelgram can be used itself as a tool to construct its own benchmark corpus with the particularity that such a corpus is either representatively exhaustive or if not exhaustive one at least knows what is missing from the corpus. The Labelgram implementation includes dynamic linguistic tracing; not only the recording of which linguistic rules are successful in disambiguating, as shown in Figure 88, but also a trace of the rules visited until a successful solution is found.

It is incumbent on the linguist to provide for each case of polycategorial ambiguity an example text exhibiting and identifying this ambiguity (for example, case '144' in Figure 88 in which 'car' is the identified ambiguous item (nom (noun) or conjonction (conjunction)). Such a text is either preferably a corpus and thus performance based text providing an attestation (together with its provenance) or, in the lack of this, a synthetic text devised by the linguist and depending thus on his competence. During the application of Labelgram to real texts, when a polycategorial ambiguity is encountered for which there is only a synthetic example, the system has the possibility to add the acquired real example as corpus based together with its provenance (this procedure being under the control of the linguist who must ensure the validity and efficacy of the 'proposed' real example), thus enriching the corpus based content of the test data set in the direction of a benchmark corpus. The component micro-systems are treated in the same way.

3.2.5   Neologisms and Jabberwocky

Labelgram can also tag and disambiguate sentences containing many neologisms. Let us take as example the first two lines from Lewis Carroll's poem Jabberwocky (Carroll 1872) containing *words* invented by Lewis Carroll but which are morphologically adequate in respect of English language morphology:

> 'Twas *brillig*, and the *slithy* toves
> Did *gyre* and *gimble* in the *wabe*;

where Labelgram's results, after expanding (normalising) "'Twas" to "It was" are shown in Figure 89.

> *'It was brillig, and the slithy toves did gyre and gimble in the wabe;'*

Our system being based on an intensional morphological algorithmic dictionary, all the *unknown words* have been automatically tagged with their part(s) of speech and these furthermore disambiguated. No other system is known which will tag properly these lines.

For applications, we know how important it is to be able to tag and disambiguate properly to ensure good results at the end. To this end, Labelgram is the first step for our Classificatim system that we now present with its associated methodology and applications.

Lexical unit	Out-of-context Categories	In-context Category
It	{PROpers}	PROpers
was	{V}	V
brillig	{ADJ}	ADJ
,	{PUNCT}	PUNCT
and	{CONJ}	CONJ
the	{ADV, DET}	DET
slithy	{ADJ}	ADJ
toves	{Nplu, V3sing}	Nplu
did	{Aux}	Aux
gyre	{V}	V
and	{CONJ}	CONJ
gimble	{V}	V
in	{ADV, ADJ, PREP}	PREP
the	{ADV, DET}	DET
wabe	{N}	N
;	{PUNCT}	PUNCT

Figure 89. Results of the disambiguated tagging of the first two lines of Lewis Carroll's Jabberwocky

Sense mining

We now turn to another methodology, itself based on our same modelling theory for language. We call the methodology in question, which has been used for major application projects over several languages and within major projects, 'sense mining'. Within sense mining, Semegram is the core system and represents our methodology.

We will start with an application system, the Classificatim system; as our theory is problem driven we will see how the theory is used and how the methodology has been developed.

3.3.1 Classificatim

Classificatim is a system which processes verbatim (Cardey et al. 2006).

What is a verbatim? A verbatim is the transcription word by word of speech. In our context due to our corpus, verbatims refer also to other items (designated this way by the industry with which we have collaborated). Verbatims are messages (contacts) from consumers; they could be:

- emails
- letters
- verbatims (transcription of telephone calls from consumers)
- etc.

The problem to be solved is that millions of messages are received every day by industries or organisations in general, and each of the messages should be analysed and classified to be then despatched to the right departments who will have to deal with the information transmitted by means of these verbatims. Our corpus came from an international agro-food industry which wanted to analyse and classify the 'verbatims' coming from their consumers. The research has been done on 7 languages and the system has to take as input a list of verbatims in French, for example (Belgium, France, and others), and produce as output the same list of verbatims as input but in addition with their associated meanings.

Classificatim is a 'verbatim' classifier. The system components are Labelgram – the parts of speech disambiguating tagger, and Semegram – a sense tagger.

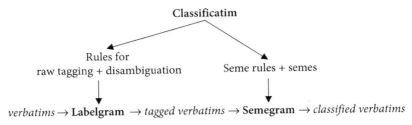

Figure 90. Representation of the Classificatim system

The Classificatim system can be represented as shown in Figure 90.
    The kernel of the Classificatim system is composed of the following:

– Labelgram: lexico-morpho-syntactic analyser that tags and disambiguates
    verbatims
– Semegram: semantico-lexico-morpho-syntactic analyser that identifies the
    semes in a given consumer verbatim using pre-established rules and semes.

The important features of the methodology are as follows:

– Easy updating – the methodology is fully traceable; problems and omissions
    can be easily pinpointed. The same approach with tracing is used as the one
    used for Labegram.
– The methodology allows disregarding certain spelling mistakes.
    For example in the verbatim:

    Power would not thicken up after adding water as instructed.
    Power is correctly recognise as *powder* even with the spelling mistake.

– The methodology allows the recognition of meanings with other representa-
    tions than keywords (e.g. syntax).
– As the work has been prepared manually, words and structures not present in
    the actual verbatims studied (corpora) have been formulated, this due to the
    linguists' intuition (competence).

Before presenting Semegram, we review the problems that have to be solved.
Imagine that we only use what are called 'keywords' or 'important words' and let
us consider the following sentence:

    the product ought to be *perfect*

perfect here does not indicate a compliment, instead it implies the understatement:
...but it is not. The consumer is really saying:

    the product ought to be perfect *but it is not*

If one tries to detect what the consumer feels using the keyword methodology with important words (*perfect*), the interpretation will be that the consumer is confident but the implied rest of the sentence which is given by *ought* says that you are probably going to lose the consumer. The interpretation is wrong if you use the said important words.

Let us take the following examples:

> For some months this product is no longer as it was before

Here who can tell what are the important words or keywords?

> The product inspires me with *confidence* and I would never have thought that I could find a product that smells.

If we take the important words here we have *confidence*, but in reality the information given by the consumer is not really this. Now for another example:

> The product would be *very good* without perfume.

Here we have the word sequence *very good*. However, what the consumer is saying is nearly the contrary.

We now present Semegram.

3.3.2   Semegram

The first thing is to define what sort of information will be relevant.

Each type of information has been organised in what we call a 'seme' or 'sub-seme'. These semes or concepts could be imprecise or indeed be very refined (we found 700 semes). Here is a well known example taken from outside the domain studied:

> If you ask for a seat, it could be answered what sort of seat do you want?
> With a back and legs (a chair)
> With a back, legs and arms (an armchair)
> Without a back, with legs (a stool)
> Without legs and without back (a cushion)

The semes and sub-semes are

> seat/stool/chair/armchair/cushion

If we take the verbatim

> The cake was not very moist very disappointed *as expect it to be as it says in the box*

the semes could be

PACKAGING/Graphics/to change/misleading/difference description-content

Semegram is organised as an algorithmic micro-system composed of rules, grouped in sets and subsets which are linked to a same seme. Semegram includes the set of semes and different sets of rules and sub-rules dealing with morphology, lexis, syntax and semantic fields. One rule represents one meaning, that is one seme, but one seme can be represented by different rules. We call such rules synonymous rules. A rule can analyse many verbatims. To enable this, we have what we call canonical formulae for representing our rules and these have an abstract representation but whose application is extensive, and being maximal thus represent in reality a great number of subformulae just like the maximal and minimal aggregates that we have already shown for the agreement of the French past participle. Take for example:

l('Find')/l('Discover')/l('Contain')/l('Have') + opt(c('Art.')) + opt(c('Adj.')) + 'piece'/'pieces' + c('Prép.') + opt(l('Colour')) + 'plastic'

This formula covers both of the following verbatims and many more:

- I found/discovered a piece of white plastic
- the product had/contained a small piece of plastic

Generality is an important factor in the system. Due to Labelgram and Semegram having been designed in intension and not in extension, the tag *verb* for example represents all the verbs in the language and *Be* represents all the conjugated forms of the English verb *to be*. We also have sets of semantic field categories which are rather like classifiers in Chinese.

Our seme classification has up to five levels. These semes could be expressed differently by consumers according to lexis and syntax. This classification has been done by hand for seven languages, these being English, French, German, Italian, Japanese, Portuguese and Spanish. We have to remark that the 'verbatims' were not always well structured and that some contained mistakes. Labelgram which is used first to tag the 'verbatims' can tag words even if they contain some mistakes (these mistakes have to be outside the morphological rules that we use for tagging) and as we have already shown, it can tag and disambiguate neologisms.

Our methodology can:

- interpret a text even if it does not contain any keywords
- analyse the full verbatim

Our methodology uses the micro-systemic methodology, mixing lexis, morphology and syntax and has already been applied on many languages. Some languages (e.g. Japanese) are classified as mostly agglutinative languages. This means that in say Japanese, units are the concatenation of 'non-empty' words and empty words. This causes a real problem with keyword methodologies that keep only 'important' non-empty words, as for example in Japanese with auxiliary verbs and non-autonomous verbs which change the verb meaning especially at the level of voice, aspect, tense and mood. This is demonstrated with the following attested examples of the verb 'Oshieru' (= inform, tell) followed by a set of non and semi-content words:

− 'Oshie' + te-hoshii = I wish that you inform me;
− 'Oshie' + te-kudasai = Please inform me;
− 'Oshie' + rareta = I was informed;
− 'Oshie' + te-ageru = I will inform (you, him...);
− 'Oshie' + te-morae-masen-de-shita = I did not get any information;
− 'Oshie' + te-morae-masen-ka = Could you kindly inform me, etc.

Languages are not made of words independent of each other; meaning can be conveyed by all sorts of structures. A preposition can play the role of a verb in Chinese for example. The order of the words could also change the meaning.

3.3.3 Testing and the classification rate

The test procedure which also enables calculating the classification rate is illustrated by the schema in Figure 91.

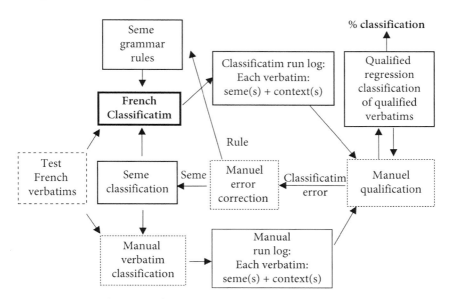

Figure 91. Testing of the Classificatim system enabling calculating the classification rate

Tests have been carried out on approximately 250,000 verbatims. The success rate with raw text (email, verbatims, and letters) and without any preparation is 84%, and after normalisation the success rate is 99%.

The manual qualification procedure for a given seme recognised in a given verbatim by the Classificatim system has already been addressed in 2.2.2.4.1 Model–driven evaluation.

3.3.4 Results over different languages

To illustrate the results over the different languages, Figure 92 shows these for verbatims containing the seme:

'*SALES & DISTRIBUTION/Availability*'

Language	Formula
	Verbatim
English	'Where' + 'can'/'could' + c('Pro.pers.') + l('buy') + l('product')/c('Pro. pers.')
	Where can I buy xxx in Cape Town?
French	'où' + opt(c('Pro. pers.')) + l('trouver') / l('procurer')
	bonjour, où trouver ce produit à Tours ?
German	'wo' +opt(Pro. pers.) + opt(Pro. pers)/ l('Produkt') + opt(Adv.) + l(bekommen)
	Können Sie mir vielleicht sagen, wo ich es noch bekommen kann?
Italian	'dove' + opt(l('potere')) + l('acquistarlo') / l('reperirlo') / l('trovarlo') / l('comprarlo')
	Vorremmo sapere se il prodotto è ancora in commercio e nel caso positivo dove potremmo acquistarlo.
Japanese	l(thm:N 製品') + (...) + 'どこ' + opt('か') + 'へ'/'に' +opt(l('/行く/')) + opt('けば')/opt('ったら')/l('と') + 1('ある')/l(/売る/')/l('見つかる') + l('/疑問/') + '#10'
	キツトカツト記念缶はどこにあるの?
Portuguese	'onde' + (...) + opt(l('poder')) + 'adquirir' / 'encontrar' / 'arranjar' / 'comprar'
	Sabe-me dizer onde é que eu poderei encontrar um produto que eu julgo que é da xxx
Spanish	'dónde' + c('Pro. pers.') + l('poder') + 'conseguir'
	no encuentro el producto y quiero saber dónde lo puedo conseguir

Figure 92. Classificatim results for the seme '*SALES & DISTRIBUTION/Availability*'

The same methodology has been used for other applications on other domains and languages such as:

- English email analysis for question answering (Zaretskaya 2012).
- Research for pertinent information in Internet texts: Arabic: (Mikati 2009) and Chinese: (Jin 2011).

The same methodology involving set based modelling and algorithmic micro-systemic analysis has been used for automatic recognition of acronyms in the context of safety critical technical documentation (Cardey et al. 2009) which is based on an intensional analysis like in the Labelgram raw tagger where no extensional dictionaries are used.

Controlled languages

Natural language abounds with subtleties and ambiguities which are all too often the sources of confusion. Such ambiguities are present at all levels: lexico-morpho-syntactic and/or phonetic. The French word '*libre*' for example has 19 senses. Moreover, several words from different languages can share a phonic identity or a graphical identity without in any way having a common meaning, or they may have a common meaning but be pronounced differently. We know too that rapid translation exists but reliable translation does not exist. Controlled languages have been created with the goal of solving problems involving readability, understanding and translation

Our theory advocates the decomposition or recomposition of linguistic phenomena in order to analyse them better. Instead of listing all the elements which make up part of a language, it is necessary rather to try to classify, order, arrange or group these elements in order to define if certain of them can function as a complete independent system or as interrelated systems, this according to what ought to be demonstrated or solved.

Confronted with a problem or a precise phenomenon, one has to choose the necessary elements and structure these in the form of a system. In this way, the problem can be manipulated because it is apprehensible (understandable). Lexis, morphology, syntax cannot be separated when one deals with language phenomena.

How does one carry out a micro-systemic linguistic analysis? It is necessary to:

- identify the problem to be treated;
- construct the system or the systems, the latter can be interrelated;
- then describe the problem using the system/s in order to solve the problem where necessary.

Based on the same theory, our methodology involving controlled languages for machine translation has been applied to domains requiring high levels of reliability in respect of machine translation as witness the LiSe (Linguistique et Sécurité) project[1] conducted in the domain of global security (Cardey et al. 2010). We start

---

1. Funded by the ANR (French National Research Agency) (Projet LiSe ANR–06–SECU–007)

with a simple example in a culinary domain of Korean translated to French (Cardey, Greenfield & Hong 2003, Cardey et al. 2005).

In French we find that the verb *arroser* in:

> *arroser le poisson de jus de citron*
> (sprinkle lemon juice on the fish)

and *saupoudrer* in:

> *saupoudrer le poisson de farine*
> (sprinkle flour on the fish)

are both translated by the same unit in Korean.

The opposition:

> [liquid/pulverulent]

has enabled marking of the difference.

The properties that we use are conceived in a relational manner. The property [liquid] is in opposition with the property [pulverulent] and this last includes, for example, the sub-properties [powdery] and [fragmented] (see Figure 93).

In this example we see that the sets are not created at the level of parts of speech or other categories that we have already seen, but this time at the level of specific features which are particularities just as were the semes in sense mining. In like manner too concerning the semes, here the actions to be described are at the level of foodstuffs. There is a micro-systemic algorithm here too, but this time this enables the best choice concerning lexis.

We will now see the same abstract micro-system applied to the domain of emergency messages and protocols that we treat within the LiSe (Linguistique et Sécurité) project (Cardey et al. 2010).

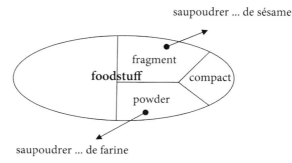

Figure 93. Micro-systemic segmenting

For example, in certain languages it is impossible to use the translation of the French word *jeter* for liquids such as water. One should use an equivalent, *verser*. Thus if we want a good translation, we should not write:

> ne pas *jeter* d'eau sur de l'huile en feu
> (do not throw water on burning oil)

Instead we should write:

> ne pas *verser* d'eau sur de l'huile en feu
> (do not pour water on burning oil)

This latter, which is just as correct in French, greatly facilitates the passage to other languages
   Thus we can write

- *Verser* QQC[liquide] sur QQC/QQN
  (pour something (liquid) on something/someone)
- *Verser* QQC[liquide] dans QQC
  (pour something (liquid) in something)
- *Jeter* QQC[solide] sur QQC/QQN
  (throw something (solid) on something/someone)
- *Jeter* QQC[solide] dans QQC
  (throw something (solid) in something)
- etc.

Intralanguage ambiguity

In every specialised language there can be units which are peculiar to the specialty and thus present no difficulties, neither for their recognition nor for their translation as for example in French law (Alsharaf et al. 2003, 2004b).

> *contre passation, crédit-bail, quasi-contrat.*

However there can be units of both the specialised and general language which can have different senses depending on their belonging to one or the other. There can also be units appearing in the specialised language with their general language senses. One can have the following partition, the domain being that of French law:

- Unit for example with a derived general use
  droit, justice, valide, arbitre;
  each term can also be a member of an expression:
  droit de légation, droit mobilier, droit moral.
- Unit of which the principal sense belongs to general language with a derived sense in the specialised language
  siège, parquet;
  that can also be found in compounds
  parquet général.
- Certain units have the same sense in both the specialty and general language:
  étude, objection, hypothèse.
- Certain units can have different senses even within the specialty:
  contrat (de gré à gré/aléatoire), force (de la chose jugée/majeure), action (coercitive/civile), libre (pratique/échange);
  this last term *libre* can have up to 19 senses (Alsharaf et al. 2003).

Again here set partitioning has been used.

MultiCoDiCT

MultiCoDiCT (<u>Multi</u>lingual <u>Co</u>llocation <u>Di</u>ctionary <u>C</u>entre <u>T</u>esnière) (Cardey, Chan & Greenfield 2006) is a dictionary which has been constructed based on sets and partitions organised in an algorithmic manner.

In respect of phenomena which concern the organisation and integrity of such a dictionary, given that such a dictionary deals with variously collocations, headwords and annotations and the various interrelations between these such as sense groups and headwords with collocations, these requirements being compounded with the dictionary being multilingual, there must necessarily be some means to ensure the integrity of these various data items and the relations between them (Greenfield 1998).

We consider linguistic phenomena and by extension lexical attributes such as annotations of linguistic phenomena as themselves a means of access to the collocations in multilingual collocation dictionaries. We illustrate this approach by describing a bilingual dictionary (French ↔ Spanish) of tourism that has been developed with this means of access in mind. Other languages have been studied too in respect of MultiCoDiCT, including Arabic, Chinese and English.

The dictionary involves the differences between French-Spanish-French translations found in ecological, fluvial and cultural (historical and religious) tourism corpora. When translating the corpora, we noticed the presence of varieties of languages (such as Latin American Spanish and the Spanish of Spain) and of regionalisms; for example, in the case of Panama whose animal and plant specimens' names were used only in the country and not in other Latin American countries, see Figure 94.

Spanish common names	Corozo	agutí
Panama	Corozo	ñeque
Bolivia	Totai	-
Cuba	-	jutía mocha
Mexico	-	cotuza
Venezuela	Corozo	zuliano de grupa negra
French translation	acrocome / coyol / noix de coyol	agouti
(English)	(coyol /coyol nut)	(agouti)

Figure 94. The presence of varieties of languages

We also found cases of linguistic phenomena, such as Americanisms, Anglicisms, non-reversible equivalents, etc. To handle these various observations we developed an algorithmic dictionary access method in order to provide access to the relevant collocations and translations. Our overall algorithm is itself composed of three principle sub-dictionaries:

a.   French-Spanish-French equivalences,
b.   particular cases at the semantic level and
c.   particular cases at the grammatical level (291 sub-conditions).

The algorithm has a maximum of eight levels of depth where the existence of other sub-dictionaries (or sub-algorithms) is possible inside each dictionary, which itself can be consulted independently or dependently. In other words, the overall algorithm includes several mini-algorithms and mini-dictionaries (see Figure 95).

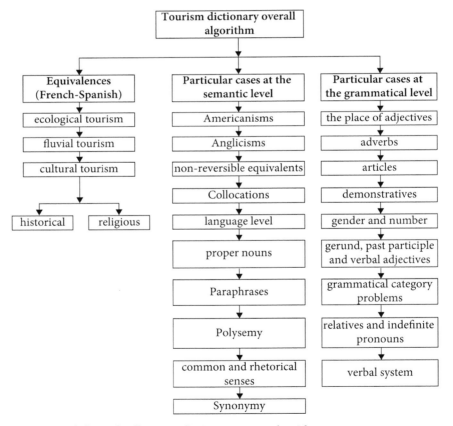

Figure 95. Multilingual collocation dictionary access algorithm

At the start of each consultation, the lexical units belonging to a given dictionary are presented in the form of a list arranged in alphabetical order so that the user can save time.

We now look at the three specific sub-dictionaries.

The first sub-dictionary concerns equivalences which are provided in the French-Spanish-French languages and which are classified according to topic. The sub-field cultural tourism presents for example historical and religious tourism as sub-sub-fields.

The second sub-dictionary concerns particular cases at the semantic field level for example, the terms of the dictionary of Panamanian fauna, for example, are joined together by class such as: insects, mammals, birds and reptiles. The user can query starting from variously:

– French to obtain the equivalences in the Spanish of Panama and common Spanish (that is the Spanish of Spain);
– French to obtain the equivalences in common Spanish and Latin;
– Panamanian Spanish to obtain the equivalences in common Spanish;
– common Spanish to obtain the equivalences in Panamanian Spanish;
– Panamanian Spanish and common Spanish to obtain the equivalences in French and Latin;
– Latin to obtain the equivalences in French and Spanish.

The third and last sub-dictionary deals with grammatical findings. It is not only composed of lexical units but also grammatical rules and also examples in order to illustrate the different cases. For this reason, we do not mention the quantity of words in the dictionary but rather the number of sub-conditions in the algorithm.

The algorithm that we have developed is interactively interpretable by the Studygram system (Cardey & Greenfield, 1992) which also provides the user interface. To show the trace of a user access using our prototype system with the dictionary access algorithm illustrated in Figure 95 we take as entry the collocation 'amazone à front rouge' where we are interested in the equivalences sub-dictionary and the particular cases sub-dictionary (see Figure 96).

A general model for access to such multilingual collocation dictionaries has been designed. The approach that we adopt involves the integration of Studygram and MultiCoDiCT. Here Studygram provides the user interface and MultiCoDiCT acts as a lexical resources server.

At one level this approach involves formalising and standardising the linguistic and lexicographic terminology shared by the two systems so as to provide a mutually agreed semantics. In any case the model underpinning MultiCoDiCT supports explicitly the concepts of collocation, sense group, synonymy, polysemy and non-reversibility of the lexis. The same model has explicitly modelled annotation

Trace of a Multilingual Collocation Dictionary
user access with as entry the French collocation
'*amazoneà front rouge*'

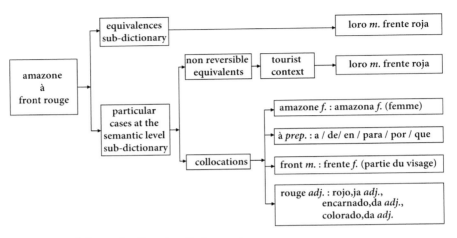

Figure 96. Multilingual collocation dictionary trace

structures attached to the sense group, to the collocation and to the collocation lexical item.

At another level this approach involves standardising the mutual call mechanism (computational operational semantics) between Studygram and MultiCoDiCT. The Studygram system in any case supports algorithm solutions (called operations) which can be procedure calls to MultiCoDiCT.

Here we again see the micro-systemic segmentation in sets and partitions, and its algorithmic implementation with canonical and variant entries which can be conditions or operators.

Controlled language and machine translation

The applications that are involved within the LiSe (Linguistique et Sécurité) project (Cardey et al. 2010) concern security in general and in particular where communication involving humans ought to be rapid and correct. Here again we need a general methodology which is based on the analysis of linguistic norms and which enables the generation of information without ambiguity, rapidly and in several languages, this being the need in the case of emergencies and crises.

When there is an international crisis, where the general population is to be warned or where protocols and medical supplies are to be transmitted or sent to a country or countries, how can we be certain that the messages, procedures or instructions are rapidly and correctly translated? Apart from rare exceptions, as one cannot predict all the messages that will be needed to be sent and translated variously to different airports for example, for warning the various populations as in the case of a tsunami, nor for the instructions to give, it is necessary to foresee the way to write the messages so that they be understood by everyone and be machine translatable without error into several languages. To achieve this it is necessary to provide the authors of such messages (which can be alerts, medical protocols or instructions of diverse nature) rules that must be respected in order to ensure readability, understanding and translatability, that is, they must be provided a controlled language. The controlled language so devised will avoid the traps of language(s) and of translation in enabling the production of source language texts with non ambiguous structures and lexical units.

Linguistic criteria have allowed segmenting and classifying controlled languages (Gavieiro-Villatte & Spaggiari 1999). Controlled languages are used in diverse domains such as aeronautics, meteorology, emergency services (police, fire, marine, ambulance etc.). English is the language the most used as the basis but one also finds Swedish and Chinese since 1988 for example. Our own objective is a double one: to enable reliable translation to several target languages and to reduce oral interferences not only within the same language but between two or more languages.

The objective of our controlled language was to enable the author, in the case of emergencies, to write a message or protocol which avoids traps concerning subsequent translation. We have carried out comparative studies between several languages in order to establish norms. What are these norms? Starting from French as the source language we have compared several French structures which, when actualised in discourse, give the same sense. We have then kept the simplest and above all the least ambiguous, this from several points of view. Firstly all the structural ambiguities have been avoided and secondly the morphological and lexical ambiguities have in their turn been replaced by recommended forms or lexical units. Equivalent structures have then been searched for in Arabic, Chinese, English and Thai and the same work has been carried out on these target languages in order to obtain equivalent structures in each of the source and target languages. As a result we have discerned 16 French structures which will thus enable the writing of all sorts of alerts and protocols, and a limited number of equivalents of these structures in the 4 target languages enabling correct translation. What we have done and which has not been done before is that we control not only the source language but also the target languages.

3.7.1   Divergent structures

The first step thus consisted in detecting what is common and what is divergent in the languages concerned. We have found structures which are sometimes completely equivalent in form and in sense, and others which are completely divergent in form but which express the same meaning. For this latter case, it was necessary to find complementary rules enabling to go from one language to the other.

We have already shown in 2.2.1.6.2 for example two syntactic structures; one in Arabic and the other Japanese. The two structures are completely divergent even though they express the same thing:

- Arabic: opt(quest +يمكن +nver) + opt(ecd) + comps/compscc/compa/compacc + point
- Japanese: comps/compscc/compa/compacc + opt(ecd + opt(quest + vinf)) + point

On the contrary, we observe that two structures (one for Arabic, the other for English) are identical:

- Arabic: opt(neg) + pred + opt(mod) + opt(compacc/compa1) + opt(compacc/compa2) + opt(compacc/compa3) + opt(mod) + opt(2pts+ liste) + point
- English: opt(neg) + pred + opt(mod) + opt(compacc/compa1) + opt(compacc/compa2) + opt(compacc/compa3)+ opt(mod) + opt(2pts + liste) + point

As we know, there are several ways to say the same thing in one and the same language. We have thus examined each language's varied and diverse structures and tried to keep those which were the most compatible with the other languages.

In this way we have been able to extract a canonical structure with its intra- and inter-language divergences.

For example (Cardey et al. 2008), if we take 'Quand le patient saigne' (When the patient bleeds) this can be translated into Chinese by means of 3 different structures, all with the same sense:

i. 一旦 + 病人 + 出血。 quand + patient + saigne
ii. 病人 + 一旦 + 出血。 patient + quand(-il) + saigne
iii. 每当 + 病人 + 出血。 chaque fois quand + patient + saigne

We can use any of these three translations. However we have chosen the first structure:

'一旦 + 病人 + 出血。 quand + patient + saigne'

because for 一旦 (quand (when)) 2 structures are possible:

a. S + 一旦 + V
b. 一旦 + S + V

where S is subject and V verb. Of these the second '一旦 + S + V' is more natural for a Sinophone. Furthermore it is more general in fitting with the Chinese SVO (O is object) structure and simplifies the transfer from the general French SVO structure.

For the sentence

> *Verser de l'eau froide sur la brûlure immédiatement durant 5 minutes*
> (Pour cold water on the burn immediately for 5 minutes)

we have for example the following canonical structures which provide good translations in the 4 languages.

– French, the source language:
   opt(neg1) + opt(neg2) + vinf + arg1 + prep_v + arg2 +
   opt(opt(prep_comp), comp1(n)) + opt(opt(prep_comp), comp2(n))

The corresponding structures for the target languages are as follows:

– English:
   opt(neg(Neg)) + vinf(Vinf) + arg1(Arg) + prep_v(Prep_v) + arg2(Arg) +
   opt(comp(Comp,N)) + opt(comp(Comp,N))

- Arabic:

  (opt)neg(Neg) + 'يجب' + nver(Nver) + arg1(Arg1) + prep_v(Prep_v) + arg2(Arg2) + opt(comp(Comp,N)) + opt(comp(Comp,N))

- Chinese:

  opt(arg0(Arg))   +   opt(neg(Neg))   +   opt(comp(Comp,N))   + opt(comp(Comp,N)) + '把'+ arg1(Arg) + v(V) + arg2(Arg)

We see again that the same methodology and the same formal representation are used as those of the preceding applications:

- Example drawn from the agreement of the French past participle:
  Pl(,)(Adv)P2(Adv)(I')(P3)(Adv)(P4)(Adv)avoir(Adj)(P3)(P4)(Adv)**List**(Adv)(Prép)(P5)Inf(P5)

- Example drawn from Sense Mining:
  l('Find')/l('Discover')/l('Contain')/l('Have') + opt(c('Art.')) + opt(c('Adj.')) + 'piece'/'pieces' + c('Prép.') + opt(l('Colour')) + 'plastic'

Formal representation items between parentheses are either singleton sets or sets with several elements. The rules are read according to a defined algorithm and according to the partitions that have been applied to the particular language being treated.

3.7.2   Lexical divergences

The same situation occurs for the lexis. One observes that different grammatical categories and specific lexical units are used for representing the same concept or seme. Equivalence tables have thus been established in order to solve divergence problems and to find for each concept an 'equivalent form' in each of the other languages. Some concept also could just not exist in some languages for example.

Certain problems such as the *dual* have emerged and are recorded and handled by equivalence tables. We note:

- Arabic: The dual refers to two individuals. It is formed from the suffix ان for the subject dual (grammatical gender) and ين for the complement dual (grammatical gender).
- Chinese: dual markers: '双' / '二' / '俩' (double)
- English: Ø
- French: Ø
- Japanese: (prefix) '双'/'二'/'両' + N
- Thai: dual markers: คู่ (double), แฝด (twins) (add these words after the noun)

As we can see, we control not only the source language but also the target language.

We will see that the super-system Ss in this context represents the controlled language machine translation system, Sc the controlled source language and Sv each controlled target language.

3.7.3 Translation architecture

Using micro-systemic linguistic analysis, we have classified and organised the equivalences and divergences in the form of a compositional micro-system structure where this is expressed in a declarative manner by means of typed container data structures and their contents so as to be incorporated in the machine translation process (Cardey et al. 2008). This has resulted in a model based on language norms and divergences with inherent tracing.

The limit 'canonical' case where there are no divergences corresponds to the target language being identical to the source language, which we call 'identity' translation. In our case this language is the French controlled language which has also been controlled for machine translation to the specific controlled target languages. Thus our controlled languages mirror each other. The architecture of our machine translation system is thus based on the variants being divergences between our controlled target languages and our canonical controlled French source language, these divergences being organised in such a manner as to effect the translations during the translation process.

The linguistic data structured by means of the typed container data structures is rooted at the root data structure named 'TACT' (Traduction Automatic Centre Tesnière) with its content shown in Figure 97.

We declare here the super-system for controlled French:

Ss_frc

Table
Tact
source_language
target_languages

source_language	
source_language	**full_form**
frC	Controlled French

target_languages	
target_language	**full_form**
arC	Controlled Arabic
enC	Controlled English
chC	Controlled Chinese
thC	Controlled Thai

Figure 97. TACT root data structure

and implicitly the 'identity' super-system for controlled and canonical French to controlled and canonical French (identity translation):

Ss_frC_frC

as well as the super-systems for controlled French to variously controlled Arabic, Chinese, English and Thai:

Ss_frC_arC
Ss_frC_enC
Ss_frC_chC
Ss_frC_thC

The typed container data structure for the canonical (source) language, controlled French, and which defines super-system Ss_frC, is shown in Figure 98. For example groups_verbs_frC is the name of a table containing the attribute name verb_fr (full name groups_verbs_frC : verb_fr) for verb lexical forms including for example associated prepositions, and the attribute name group_frC for the French group to which the verb / verb + preposition is associated. The table groups_frC contains the controlled classification that we have undertaken of the French verbs' syntactic structures organised as a micro-system.

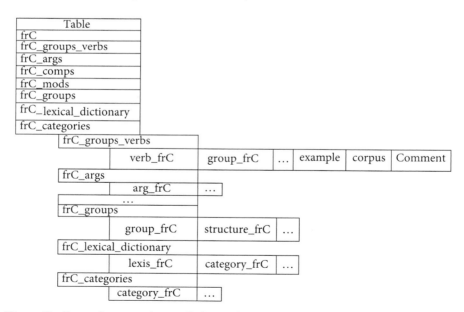

Figure 98. Source language (controlled French) container data structure defining super-system Ss_frC

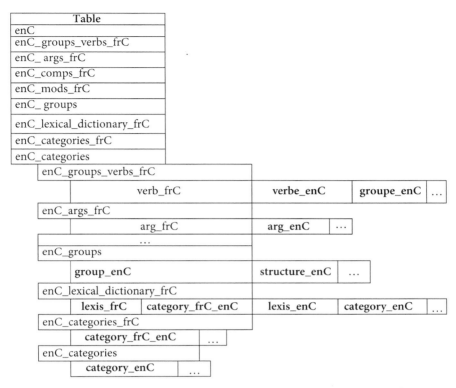

Figure 99. Target language (English) data structure defining super-system Ss_frC_enC

The typed container data structure for one of the variant (target) languages, controlled English and which defines super-system Ss_frC_enC, is shown in Figure 99. In the figures illustrating the container data structures, we highlight any divergences with the previous container data structure figure in bold, for example **verbe_enC** and **category_frC_enC**.

At the outset it is important to observe that this super-system (Figure 99) is indeed modelling the relation between controlled French and controlled English. Thus it will be seen that certain table names correspond to their controlled French counterparts but with divergence (i.e. variant) indications. For example canonical source French table name groups_verbs_frC corresponds to variant target English table name enC_groups_verbs_frC. As to the tables' contents, these are respectfully (canonical French) verb_frC, group_frC and (variant English) verb_frC, verbe_enC, groupe_enC.

A datum in enC_lexical_dictionary_frC: category_frC_enC which is unset means that for the English translation of the French lexical item the French grammatical category applies; otherwise enC_categories_frC: category_frC_enC contains a datum. This latter is the case for example for the French lexical unit 'maison' which has the French category nfs (in categories_frC: category_frC) but ns in enC_categories_frC: category_frC_enC. What is being illustrated here is that the grammatical category of 'maison' is a variant for translation to English. The French lexical unit 'la' however has an empty datum in enC_categories_frC: category_frC_enC meaning that for translation, the canonical French category in categories_frC: category_frC applies (adfs). We see by means of this illustration that not only are the typed container data structures organised in canonical and variant fashion, but so too are the linguistic data that they contain.

It is legitimate to ask concerning the definition of the 'identity' super-system for canonical translation of controlled and canonical French to controlled and canonical French, Ss_frC_frC. This super-system is a concrete micro-system which is used for example as a control in the verification of the implemented system where controlled French is indeed mechanically translated to controlled French. In fact the definition of Ss_frC_frC is implicit and its definition involving its instantiation is carried out mechanically during system initialisation.

The typed container data structure for the variant (target) language, controlled Thai, and which defines super-system Ss_frC_thC is shown in Figure 100. It will be observed that this is identical to that of English (Figure 99), and indeed to that for Ss_frC_frC, apart from the renaming of the table and attribute names (en → th). The '[=]' indicates that all the linguistic data table structures are identical except for the renaming (going from groups_verbs_frC_thC to categories_thC).

The typed container data structure for controlled Arabic, and which defines super-system Ss_frC_arC, as one of the variant (target) controlled languages is shown in Figure 101. This typed container data structure is identical to that for canonical Ss_frC_frC as well as of English and Thai (Figures 99 and 100) apart from

Table
thC
thC_groups_verbs_frC
thC_args_frC
thC_comps_frC
thC_mods_frC
thC_groups
thC_lexical_dictionary_frC
thC_categories_frC
thC_categories

[=]

Figure 100. Target language (Thai) data structure defining super-system Ss_frC_thC

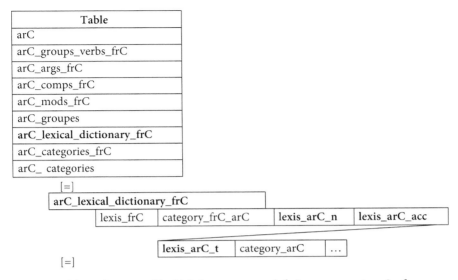

Figure 101. Target language (Arabic) data structure defining super-system Ss_frc_arc

the table and attribute names (en → ar), with the exception of arC_lexical_dictionary_frC where we see the three Arabic attributes names lexis_arC_n, lexis_arC_acc, lexis_arC_t corresponding to the single English attribute name lexis_enC. Linguistically, arC_lexical_dictionary_frC accommodates the Arabic morphological forms nominative, accusative and tanwin. In this respect, in terms of the target languages, Arabic is the exception, whilst English, Chinese and Thai are 'regular'.

Figure 102 shows the typed container data structure and which defines super-system Ss_frC_chC for the variant (target) language, controlled Chinese. Here there are three differences with that for canonical Ss_frC_frC and for English and Thai apart from table and attribute names. The first concerns table chC_groups_verbs_frC with the extra attribute indicator_chC. This attribute's values are used to select the relevant indicator within the linguistic data in chC_groupes: structure_chC. The second difference, also concerning table chC_groups_verbs_frC involves the French verb/verb + preposition groups which are used for the other controlled target languages and which are condensed for Chinese; this table contains the necessary reformulations and the mappings to the French verb/verb + preposition groups. The final difference concerns Chinese classifiers; see chC_lexical_dictionary_frC: classifier_chC.

We now turn to the scalability aspects of our methodology. This concerns two dimensions:

1. the extension to other target languages;
2. the possibility of incorporating other source languages.

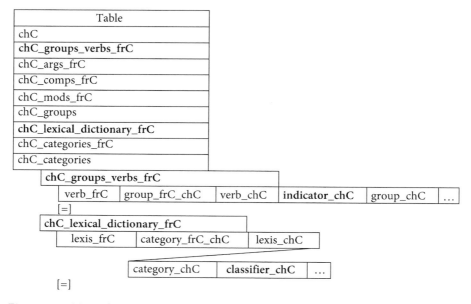

Figure 102. Target language (Chinese) data structure defining super-system Ss_frc_chc

In respect of the first dimension involving extending to other target languages our architecture is conducive to facilitating this endeavour; the costs are clearly visible. For example Arabic with its complex morpho-syntax necessitates the formulation of associated post transfer processing neither shared with Chinese nor Thai. On the other hand, the Arabic model has been extended to Russian (Jin & Khatseyeva 2012), the grammatical case micro-system being ready for this class of language.

As to the second dimension concerning the incorporation of other source languages, there are two questions concerning:

1. recuperation of data from our single source language controlled French;
2. recuperation of data where a new source language is an existing target language.

For the first of these questions, we have already worked with Korean as a source language (Cardey, Greenfield & Hong 2003) where it can be seen that the approach using the verb as the root of the analysis is the same as we have in the LiSe project.

In respect of the second question, recuperation of target language data, we observe that the canonical translation (source and target languages being the same and thus reversible) can serve as a baseline for attempting to 'extract' a source language from an existing target language whatever the target language be.

We also observe that it would be interesting to investigate language level operations on divergences as a basis for exploring these two dimensions.

3.7.4   Including a new language pair: Russian to Chinese

TACT, our machine translation system, is a novel hybrid (pivot + transfer) rule-based machine translation architecture in which the pivot language (PL) is French controlled also for translation, and where the Transfer System is directed by the various source-target language divergences.

A great advantage is that our system architecture is modelled in a way to ensure that it remains fully extendable and thus requires minimal changes when adding a new target language. At the same time it does need a complete new linguistic module, containing a dictionary and a complete micro-systemic representation of the new target language. So, once our system is prepared for adding the new target language, we can proceed to the formalisation of our controlled Russian to Chinese so as to design its linguistic model.

The transfer super-system operates first between controlled source French and controlled pivot French, and then between controlled pivot French and controlled target Russian/controlled target Chinese. The divergences are incorporated to our system and connections are made at the 'super'-level.

If we take as example the French sentence:

Mettre chaque manette de poussée en position IDLE immédiatement.
(Put each thrust lever in IDLE position immediately)

This French sentence will be represented in our formalisation language in the following form:

opt(neg1) + opt(neg2) + vinf('mettre') + arg1 + prep_v('en') + arg2 + opt(prep_comp) + opt(comp, Na) + opt(prep_comp) + opt(comp, Na)

We see here the correspondences between our French verb and target verb structures in Chinese:

indicateur_chC('在'/' 把') + arg1 + v + arg2

and in Russian:

v + arg1_acc + prep_v('в') +arg2_acc

3.7.5 Tests

We have compared our machine translation system TACT with others.

Take the sentence:

> *Ne pas brancher trop d'appareils électriques sur la même prise à la maison*
> (Do not connect too many electrical appliances to the same plug at home.)

The results obtained are as follows:

– Systran and WorldLingo
  Not to connect too many apparatuses on the same catch at the house
  不连接许多electricals器具在同样捉住在房子
  مسكت لا أن يربط [توو مني] [إلكتريكلس] تطبيق على ال نفس في المنزل
  Не подключить too many прибора electricals на этих же уловите на доме
– SDL FreeTranslation
  Do not connect electric too many devices on the even taken one to the house
– Reverso
  Do not connect too many devices on the same grip(taking) to the house
– Google
  Do not plug too many appliances on the same prize at home
  نفس الجائزة في المنزل لا سد العديد من الأجهزة على
  不要太多外挂设备在同一奖项在家里
  Не подключайте слишком много приборов в тот же сокет дома.
– TACT (our system)
  Do not connect too many electrical appliances to the same plug at home
  لا يجب وصل الكثير من الأجهزة الكهربائية بـنفس المنشب الكهربائي في المنزل
  不要在家把太多的电器插在同一个插销。’
  อย่าเสียบปลั๊กเครื่องใช้ไฟฟ้าจำนวนมากเกินไปบนที่เสียบอันเดียวกันที่บ้าน'
  Не подключать слишком много электрических приборов в одну и ту же розетку дома.

The results are good for us as the system has been designed according to our theory and methodology going from problem solving to the way to do it.

3.7.6 Tracing

What is interesting is that with our system we also obtain a trace.

Firstly the author composes messages syntagmatically in the controlled French pivot language using the LiSe User Interface (Cardey 2008b). The two traces that we show start from this syntagmatically segmented source sentence.

We give as an example the trace of the translation of the same controlled French sentence:

Ne pas brancher trop d'appareils électriques sur la même prise à la maison

to English (Figure 103) and to Arabic (Figure 104).

```
| ?- test_traduire_langue(1,enC, Traduction).
RegroupageEnSyntagms_LS =
    [['Ne',neg1],[pas,neg2],[brancher,vinf],['trop d'appareils
    électriques',arg1],[sur,prep_v],['la même
    prise',arg2],[comp(lieu,'à la
    maison'),comp1],[comp('',''),comp2],['.',pt]]
LS = frC, LC = enC
LC_GroupeVerbal_LS =
    [brancher,frC_7,'',qqc,'',sur,qqc,'','','','','','','',connect,'enC_
    1.7','',to,prep_v,'','','','','','Incendie','']
Regroupage_En_Arguments_LS =
    [neg1 - [['Ne',neg1,'Do',neg1,'','Incendie','']],neg2 -
    [[pas,neg2,not,neg2,'','Incendie','']],vinf -
    [[brancher,frC_7,connect,'enC_1.7']],arg1 - [['trop d'',det,'too
    many',det,'','Incendie','],[appareils,nmp,appliances,np,'','Ince
    ndie','],[électriques,adjmp,electrical,adj,'','Incendie','']],pr
    ep_v - [[sur,prep_v,to,prep_v]],arg2 -
    [[la,adfs,the,ad,'','Incendie','],[même,adjs,same,adj,'','Incend
    ie','],[prise,nfs,plug,ns,'','Incendie','']],comp1 -
    comp(lieu,[['à la maison',adv,'at home',adv,
    '','Incendie','']]),comp2 - comp('',[]),pt -
    [['.',pt,'.',pt,'','','']]]
Unites_Source = frC
    ['Ne',pas,brancher,'trop
    d'',appareils,électriques,sur,la,même,prise,'à la maison','.']
Regroupage_En_Arguments_LC =
    [neg1 - [['Ne',neg1,'Do',neg1,'','Incendie','']],neg2 -
    [[pas,neg2,not,neg2,'','Incendie','']],vinf -
    [[brancher,frC_7,connect,'enC_1.7']],arg1 - [['trop d'',det,'too
    many',det,'','Incendie','],[électriques,adjmp,electrical,adj,'',
    'Incendie','],[appareils,nmp,appliances,np,'','Incendie','']],pr
    ep_v - [[sur,prep_v,to,prep_v]],arg2 -
    [[la,adfs,the,ad,'','Incendie','],[même,adjs,same,adj,'','Incend
    ie','],[prise,nfs,plug,ns,'','Incendie','']],comp1 - [['à la
    maison',adv,'at home',adv,'','Incendie','']],comp2 - [],pt -
    [['.',pt,'.',pt,'','','']]]
Unites_Cible = enC
    ['Do',not,connect,'too
    many',electrical,appliances,to,the,same,plug,'at home','.']
Traduction =
    'Do not connect too many electrical appliances to the same plug
    at home.' ;
```

Figure 103. TACT system: trace of the translation of the controlled French sentence *Ne pas brancher trop d'appareils électriques sur la même prise à la maison* to English

```
| ?- test_traduire_langue(1,arC, Traduction).
RegroupageEnSyntagms_LS = Identical to French to English
Regroupage_En_Arguments_LS =
[neg1 - [[ne,neg1,لا,'','','','','','',neg1,'','Incendie','']],neg2 -
   [[pas,neg2,'','','','','','','',neg2,'','','']],

   [...]

   comp1 -
   (([[à,prep_comp,'','','','','','','',''],[la,adfs,ال,'','','','','','',a
   rtu,'','Incendie',''],[maison,nfs,'',منزل,'','منزل,منزلا,منزل,منزل,منزل,منزل,nms
   ,'','Incendie','']] -
   [[prep_comp(à),ad,nfs],[prep_comp(في),artu_,n(acc)]],'à la maison'
   = 'في ال منزل','Incendie','']),comp2 - ([] - []),pt -
   [['.',pt,'.','','','','','',pt,'','Incendie','']]]
Unites_Source = frC
[ne - neg1,pas - neg2,brancher - frC_5,'trop d'' - det,appareils -
   nmp,électriques - adjfp,sur - prep_v,la - adfs,même - det,prise -
   nfs,à - adfs,maison - nfs,'.' - pt]
Regroupage_En_Arguments_LC =
[neg1 - [[ne,neg1,لا,'','','','','','',neg1,'','Incendie','']],lexis -
   [['','',يجب,lexis]],neg2 -
   [[pas,neg2,'','','','','','','',neg2,'','','']],nver -
   [[brancher,frC_5,وصل,'arC_5.1']],arg1 - (([['trop d'',det,' الكثير
من
',detv],['','',ال,lexis_],[appareils,nmp,أجهزة,nfs],['','',ال,lex
is_],[électriques,adjfp,[n : nfs > كهربائية,n : nms > كهربائي,n :
nmp > كهربائيين,n : nfp > كهربائيات,n : nfd > كهربائيين,n : nmd >
كهربائيين],adj]] -
   [[det,n,adj],[detv,lexis_(ال),n(acc),lexis_(ال),adj(acc)]]),prep_
v - [[sur,prep_v,ب_,prep_v_]],arg2 -
   (([[même,det,نفس,detv],[la,adfs,ال,artu_],[prise,nfs,' منشب
الكهربائي',nms]] - [[ad,det,n],[detv,artu_,n]]],compl -
   (([['','',في,prep_comp],[la,adfs,ال,artu_],[maison,nfs,منزل,nms]] -
   [[prep_comp(à),ad,nfs],[prep_comp(في),artu_,n(acc)]],'à la maison'
   = 'في ال منزل','Incendie','']),comp2 - ([] - _6411784),pt -
   [['.',pt,'.','','','','','',pt,'','Incendie','']]]
government_agreement('Governor > Governed'(n : nfs >
   كهربائية,(كهربائية),'Lexique_LC'(كهربائية),'Arg_LC'([detv,lexis_(ال),n(acc)
   ,lexis_(ال),adj(acc)]),'Governed_ArgElement_LC'(adj(acc)))
Unites_Cible = arC
[يجب,'' - لا - lexis,'' - '',وصل - 'arC_5.1','الكثير من ' - detv,ال -
   lexis_,أجهزة - nfs,ال - lexis_,كهربائية - adj,ب_ - prep_v_,نفس -
   detv,ال - artu_,منشب الكهربائي' - nms,في - prep_comp,ال -
   artu_,منزل - nms,'.' - '']
Traduction = ' لا يجب وصل الكثير من الأجهزة الكهربائية بـــنفس المنشب
   الكهربائي في المنزل.'
```

Figure 104. TACT system: trace of the translation of the controlled French sentence *Ne pas brancher trop d'appareils électriques sur la même prise à la maison* to Arabic

In both cases, for targets English (Figure 103) and Arabic (Figure 104), the source sentence structure segmented at the syntagmatic level, `RegroupageEnSyntagms_LS`, is necessarily identical. However `Regroupage_En_Arguments_LS`, which is the source sentence structure with its arguments values, is realised according to the target language. For example, for English as target language, source 'à la maison', elements of the phenomenon indicated for example thus '**à la maison**', is tagged (see `Regroupage_En_Arguments_LS`) as a single French adverb corresponding to the English adverbial locution 'at home', whilst for Arabic as the target language, 'à la maison', is structured as a sequence of French individually tagged lexical units.

Looking further down in the trace for French to Arabic (Figure 104) we have highlighted phenomena which are absent from English but which must be handled for Arabic. These phenomena are variously (1) noun-adjective agreement indicated *thus*, (2) case marking (accusative) indicated **acc**, and (3) clitic processing. For clitic processing, starting at `Regroupage_En_Arguments_LC`, at the level of the formalisation, the prefixing clitic indicators, which are shown thus '_', are coded as postfixed to the relevant prefixing clitic categories, such as `artu`, and are shown thus '`artu`_'. The cliticisations in `Traduction` are shown, for example, 'المـنـشب' thus, remembering that Arabic is written from right to left.

We have seen methodologies for written language but let us now look at oral ambiguities which also present problems.

CHAPTER 3.8

Oral

Because one does not expect any individual to master and speak a number of languages with the same level of competence as that reached in their own mother tongue, industry prefers a precise and concise language to the use of a non-controlled natural language which could allow ambiguities of various types.

3.8.1   Controlling the oral

We have already mentioned that ambiguities are found at all levels: morphological, lexical, syntactic, and/or phonetic. Many words of different languages can share a phonic or graphic identity, without however having the same sense, or have the same sense but be pronounced differently. Take for example:

>   *passe* (French), and *pass* (English), e.g. an examination,

the French word means 'take' whilst the English means 'succeed' when on the contrary

>   *cul de sac*

has the same meaning in French and English, but the pronunciation is different. Thus the second important requirement of our controlled language is that it should not allow any inter-language or intra-language oral ambiguity.

We now look at the problems that we have handled.

The lack of differentiation between certain phonemes for the speakers and listeners can provoke confusion which can sometimes have grave consequences. The problem is that a message can be pronounced with diverse accents and listened to by persons who do not necessarily have the same phonetic reference system in their respective languages. Take for example the English consonant phoneme opposition system shown in Figure 105 (we use the SAMPA (phonetic system)). English was of interest to us within one of our projects involving aeronautics as the language used is American English (AECMA. 1995, Spaggiari et al. 2005).

p	t	k	f	T	s	S	tS
b	d	g	v	D	z	Z	dZ

Figure 105. Consonant phoneme opposition system

3.8.1.1   Quasi-homophones and recognition

In her PhD dissertation (Gavieiro-Villate 2001), E. Gavieiro-Villatte pointed out the phenomenon of those words in the same language which have a graphical form so close that they could be confused, and more particularly so when in stressful situations. There is the case for example of words which differ only in just one of their letters or phonemes. We have thus not only addressed the problem of avoiding the use of homophones and homographs in the same language but also the use of those which can be called quasi-homophones and quasi-homographs. We can define these in the following manner: units are quasi-homophones or quasi-homographs when only one or two of their constituents (here letters and phonemes) are different. This means that one of them can have:

- – one less or one more phoneme
  (*aft-after*),
- – a different phoneme
  (*check-deck, feed-feet*),
- – or even two different phonemes
  (*flap-slat*).

Thus we ought, amongst others, to recommend against using pairs of this type in the same language or in different languages where there is a difference only in one phoneme or two when they are pronounced.

For example one finds

the opposition *d/t* in *feed* and *feet*.

Phonemes which occupy the same position in words can also be confounded as

check and *deck*, *flap* and *slap*, and for French *dessus* and *dessous*.

We take another example in French – in fact this type of problem appears in all languages – if one says to an air pilot to:

perdre de l'altitude (perdre = lose),

it is important that he does not understand this as

prendre de l'altitude (prendre = take).

The units *altitude/attitude/latitude* present the same type of confusion and above all the letters are very close to each other visually. This problem is not surprising if one knows that reading is not linear.

Reading

One understands the problems just cited better in reading the two texts in Figure 106; in fact anyone capable of reading English or French can read these texts (the phenomenon being a research topic at the MRC Cognition and Brain Sciences Unit, Cambridge, UK).

This is yet another reason for which by means of controlled languages one ought to avoid using lexical units which present the same letters or phonemes as other lexical units.

> it deosn't mttaer in waht oredr the ltteers in a wrod are, the olny iprmoetnt tihng is taht the frist and lsat ltteer be at the rghit pclae. The rset can be a total mses and you can sitll raed it wouthit a porbelm. Tihs is bcuseae the huamn mnid deos not raed ervey lteter by istlef, but the …
>
> l'odrre des ltteers dnas les mtos n'a pas d'ipmrotncae, la suele coshe ipmrotnate est que la pmeirère et la drenèire soit à la bnnoe pclae. Le rsete peut êrte dnas un dsérorde ttoal et vuos puoevz tujoruos lrie snas porlbème. C'est prace que le creaveu hmauin ne lit pas chuaqe ltetre elle-mmêe, mias …

Figure 106. it deosn't mttaer & l'odrre des ltteers

3.8.1.2 Generation and interpretation of language sounds

Bad pronunciation can generate homophones and change the sense of a message. We show in what follows some problems that an approximate pronunciation can engender. We take the case (Figure 107) of English words pronounced by a Thai and the ambiguities which are produced, knowing that the following phonemes do not exist in Thai:

[T], [D], [g], [v], [z].

English phonemes	Sound in Thai	Ambiguities
[T]	[t], [d]	birth → bird
[D]	[d]	they → day
[v]	[w]	vine → wine

Figure 107. Ambiguities resulting from the absence of certain English phonemes in the Thai language

We will not go further in terms of detail here but we can already note that *vine* for example confounds with *wine*.

Application

In 1998, Airbus Operations SAS started a project, in which we participate, dedicated to the creation of a controlled language for industrial use; in this case enhancing warning text quality in aircraft cockpits. Another objective was to provide designers with a means of facilitating their job whilst respecting stringent safety criteria.

Methodology

Firstly, in working with linguists with different mother tongues, we have attempted to see which English phonemes do not belong to their own language.

Secondly, we have tried to see with which phonemes a native who is not specialist in English will tend to replace the non existing phoneme.

Thirdly we have shown that these pronunciations could cause ambiguities in English.

3.8.1.3   Examples of 12 languages with problems at the level of phonemes compared with English

We have here an example of interlingual 'divergent equivalences'.

Sets of phonemes to be used and sets of phonemes to be avoided were constructed which means that sets of lexical units containing certain phonemes to be avoided have also to be created as well as sets of units which can replace the lexical units which are not allowed.

We now present examples in 12 languages with problems at the level of phonemes compared with English, (in fact 13 as we grouped Norwegian and Swedish together).

Arabic (Lebanese)

In Arabic, the following 2 sounds do not exist:

> [v] as in 'video'
> [p] as in 'people'

In fact:

> [v] is pronounced [f]
> [p] is pronounced [b]

Ambiguities:

> pack   →   back
> vox   →   fox

[g] as in 'gun' does not exist in Arabic but can be pronounced by Saudi Arabians, Palestinians and Egyptians, who have a tendency in their own language to replace the uvular consonant [q] by [g].

Other Arabic populations (Syrians for example) who do not use [g] will replace it by [k]; for example: goal → coal

Chinese

Sounds which do not exist in Chinese: [T], [D], [v], [z]

> [T] is replaced by [s]:   path faith myth
> [D] is replaced by [d]:   they though
> [v] is replaced by [b]:   value van

For these 3 sounds Chinese is like Korean.

> [z] is generally well pronounced (some Chinese could pronounce it like a plosive).

For other cases the replacement could vary.

Sounds which exist in Mandarin Chinese but that Taiwanese do not pronounce correctly because of the dialect (Taiwanese or Minnan spoken in the south of China) and because of the influence of Japanese (the Japanese occupation of Taiwan 1895–1945):

> [r] (at the beginning of the syllable) is replaced by [l]: road → load

In extreme cases (especially in the south of China):

> [l] (at the beginning of a syllable) is replaced by [n]: line → nine

As far as vowels are concerned, some Chinese tend to pronounce [{] like [E]; in general it rarely creates ambiguities (la̱st → le̱st).

Examples of ambiguities which could appear

pa̱th	→	pass	/	fai̱th	→	fa̱ce	/	my̱th	→	mi̱ss
the̱y	→	da̱y	/	tho̱ugh	→	do̱ugh	/	va̱n	→	ba̱n
ro̱ad	→	lo̱ad	/	li̱ne	→	ni̱ne	/	la̱st	→	le̱st

Dutch

Sounds which do not exist in Dutch:

[D], [T], [S], [tS] (the last 2 do not cause real problems)

One same phoneme could be replaced by more than one in the other language:

[T] as in 'thi̱nk' could be replaced by [f], but in 'bo̱th' for example we never hear [f] instead of [T].

Sometimes when [T] is at the end of the word it could be pronounced [t].

[D] is often replaced by [d], especially when the English word looks like the Dutch one:

'mo̱ther'	→	'moe̱der'
'thi̱s'	→	'di̱t'
'Ne̱therlands'	→	'Ne̱derland'

Ambiguities:

fai̱th	→	fa̱te
the̱y	→	da̱y

Japanese

Sounds which do not exist in Japanese:

[T], [D], [f], [v],[V],[{], opposition [r]/[l]

We find:

[T] is often replaced by [s]:	pa̱th fai̱th my̱th
[D] is pronounced [z]:	the̱y tho̱ugh
[f] is replaced by [h(w)]:	fi̱ve fe̱el
[v] is replaced by [b]:	va̱lue va̱n
[{] is replaced by [a]:	ra̱n, ta̱bloid
[r] is replaced by [l]:	ro̱ugh ra̱re

Ambiguities:

| path | → | pass | / | faith | → | face | / | myth | → | miss |
| van | → | ban | / | rough | → | laugh | | | | |

Korean

Sounds which do not exist in Korean:

[T], [D], [f], [v]

We find:

[T] is replaced by [s]:   path faith myth
[D] is replaced by [d]:   they though
[f] is replaced by [p]:   five feel
[v] is replaced [b]:   value van

Ambiguities:

path	→	pass	/	faith	→	face	/	myth	→	miss
they	→	day	/	though	→	dough				
fan	→	pan	/	van	→	ban				

Malay

Change of consonant:

[f] is often replaced by [p]
[D] is often replaced by [d]
[S] is pronounced [s]

Suppression of the non existing consonant:

[r] germ → gem (in Malay the sound [r] is always followed by a vowel)
[S] scotch → scot

Norwegian/Swedish

then	→	den
thick	→	tick
grows	→	gross
chins	→	chintz
zoo	→	sue

Other non-Scandinavian phonemes that can easily be mispronounced are as follows:

badge → batch
jelly → yelly
while → vile
wine → vine

Persian

[w] as in 'war' is replace by [v].

Persians add a vowel in front of consonants at the beginning of words:

student → estudent

Polish

Diphthongs do not exist in Polish. The juxtaposition of vowels seems to be sufficient.

[T] is pronounced [f]
[D] is replaced by [d]

Ambiguities:

they → day / though → dough
think → fink

Romanian

[T] is replaced by [s]:   thin, thing, thick
[T] is replaced by [t]:   thumb, thanks (thank you), heath
[D] is replaced by [d]:   thy, they

Ambiguities:

thin → sin / thing → sing / thick → sick
thanks → tanks / heath → heat
thy → die / they → day

Spanish (Latin American)

[D] is replaced by [d]
[v] is replaced by [b] or [B] (from Spanish)

We find vest → best

[z] is replaced by [s]
[{] is replaced by [a]

We also find:

juice is pronounced with [j] like in use
yes and Jess are both pronounced / jEs /

Remarks

We have noticed that the recurrent English phonemes which are missing in other languages are [T] and [D]. A complete table showing the interferences on the languages that we have studied has been drawn up and serves for the creation of the controlled language. What complicates things is that those phonemes approximately or incorrectly pronounced could be interpreted by another phoneme according to the mother tongue/nationality of the hearer (Cardey 2008a). We have then to provide rules so that this sort of ambiguity does not appear in the lexis of the controlled language.

Sets and rules have been created as well as an algorithm to decide what to use and when.

3.8.1.4   Other problems for future work

We know that 'standard' languages cohabit with different levels of language (literary, familiar and so on) and that they are used by different social classes; to these we have to add regionalisms as illustrated by regional dialects. Often people use words without knowing that they just belong to their specialised language, or to their region.

Languages are in contact with one another and thus influence each other in different ways, especially by the borrowing of words. English has borrowed from more than 350 languages over the centuries. Today the phenomenon is inversed, and languages borrow a large number of words from English, the international language for exchanges.

Another problem is that all languages do not use the same writing system.

Some research has already been done concerning the problem of imported words from languages to others but this research has to be enlarged to many more languages and must become more general and more refined.

We have seen why language interferences have to be avoided and we have proposed some ways to find and solve these problems. A lot of work still has to be done but the methodology which consists in working at the level of phonemes instead of trying to find individual words seems to be more productive and easier to generalise for solving the problems of interferences which are due to bad pronunciation or bad interpretation. The methodology used allows tracing back to the cause of the problems which is essential in safety critical applications.

We have shown in this final part of the book applications based on our theory using sets, partitions and relations, all working in intension and processed by micro-systemic algorithms.

Conclusion

To conclude, in our book we have tried to show how to model language, a concept which is materialised in different languages of different origins which present resemblances but also numerous divergences.

Our comparison with galaxies, stars, aggregates and elements has enabled us to show using our micro-systemic linguistic theory, perhaps in a clearer way, how to model the organisation of these languages.

We have not only given our own point of view concerning considering languages as systems, but also those of grammarians and linguists over very many centuries. Problems, much due to terminology, arising from the meta-language used for describing languages, as well as to representations, have been exposed so as to understand better our theory's theoretical foundations. We have briefly described those elements needed for the theory's construction, and also for the student reader we have provided reminders of certain fundamental notions.

In reality, the systemic approach that we have mentioned here is based on logic, set theory, partitions and relations, and also the theory of algorithms. The easy and representative example of the doubling of the final consonant in English helps in understanding these various notions and also shows that syntax, morphology, phonology, lexis and even register are unified in the same systemic representation which leads to the algorithm indicating how to solve the problem.

A theoretical approach which is mathematically based, whatever it is, ought to be able to accept linguistic formalisms. For this reason such an approach has to be sufficiently flexible so as to enable the construction of models themselves founded on model theory using a constructive logic approach. These models must adapt themselves as the analysis proceeds and when new problems are uncovered. Our linguistic approach involves the delimitation of sets by choosing only and uniquely those elements which serve to solve the problem, the sets so involved being able to function together in terms of relations. In order to describe a language, nothing ought to be fixed in advance, the risk being to find oneself tied into a situation where instead of analysing the language, one bends and deforms the said language and its analysis to fit the supposed model, such a practice being not only against nature, but anti-productive, anti-descriptive, and contrary to and the inverse of all that is theoretical and experimental scientific practice.

Whilst systemic linguistics does not include a computational aspect per-se, and furthermore has been applied for applications from teaching grammar to safety critical controlled languages, the nature of the analyses is conducive to computational processing. As well as for end user applications, such computational processing can be useful, for instance, for the mechanical verification of linguistic data, for grammatical concordances and traceability and also for automated case based benchmark construction, etc.

To conclude concerning the methodologies and applications, we are of the opinion that dictionaries ought to be created according to the needs. Their form, content and their manner of access should vary according to what one wants to do.

We would like also to elucidate the difference between Sense Mining and Machine Translation. We now are aware that extraction and generation require the same type of information and that it is thus possible to use the same methodology for machine translations and sense mining. What then is the relation between the two of them? If we look closely at extracting sense and translating, which means generating sense, we see that both share the same model but that their respective processes are reversed. When translating, we start with a surface structure to which a meaning is attached. This meaning can be paraphrased in different surface representations not only in the source language but also in the target language(s). On the contrary, when we search for how a meaning is represented, as in sense (seme) mining, we start with a meaning and look in texts for its surface representations. It is interesting to see how such a relation between machine translation and sense mining can be applied, on what domains, and, above all, how the relation is transferable not only from one domain to another, but also from one language to another, however different these languages may be.

In the same manner as our opposition Sense Mining – Machine Translation, a similar opposition is present between Controlled Languages and Text Normalisation; and so too between Text Normalisation and Language Register Mining. Likewise we have complementarities rather than oppositions, as between Controlled Languages and Machine Translation, and Text Normalisation before Sense Mining, and indeed Text Normalisation as the standard 'canonicaliser' for open language interoperability...

In sum, with our theory we see unfolding before our eyes relations whether oppositions or complements for potential applications that we had never imagined.

We have chosen safety/security critical applications which require fault-free systems. It goes without saying that the two interrelated problems with which we are confronted are those of polysemy and homonymy at all the interrelated levels of syntax, lexis and morphology and we have seen that the same problem occurs at the oral level.

Epilogue

As we started with our galaxy let us finish with a star representing the sets, partitions, relations and micro-algorithmic theory of a morphological problem which we show in Figure 108, that of the plural of the French adjective. The star representation is read as follows:

Start at Adjectives 1. Go to "hébreu". If yes add "X". If no, go to "bel, nouvel, vieil". ... If no, go to onomatopoeia etc. If yes go to 2. If yes apply the operator. If no apply the main operator "does not change".

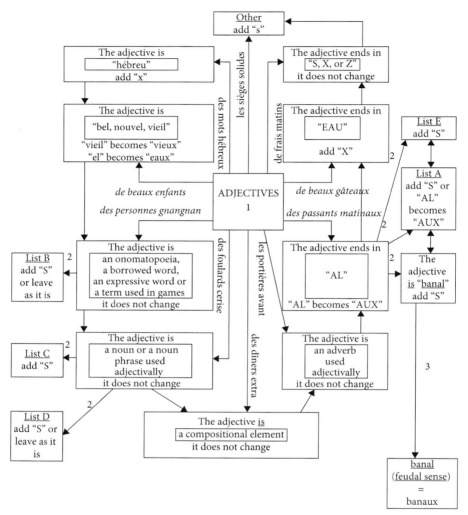

Figure 108. From the Galaxy to the Algorithmic Micro-system as a Star

List A: austral, boréal, estival, final, frugal, génial, glacial, idéal, jovial,
marial, natal, pascal, pénal
List B: bougon, chic, grognon, impromptu, mastoc, rococo, snob, souillon
List C: écarlate, fauve, géant, mauve, pourpre, rose
List D: angora, châtain, kaki, monstre, orange, pie
List E: bancal, causal, fatal, naval, nymphal, tonal, tribal

References

AECMA. 1995. *Simplified English. Association Européenne des Constructeurs de Matériel Aérospatial. A guide for the preparation of aircraft maintenance documentation in the international aerospace maintenance language*, Bruxelles: Gulledelle.

Aho A.V., Ullman, J.D. 1995. *Foundations of Computer Science C Edition*. New York: Computer Science Press,

Alsharaf H., Cardey, S., Greenfield, P. 2004a. French to Arabic Machine Translation: the Specificity of Language Couples. In *Proceedings of The European Association for Machine Translation (EAMT) Ninth Workshop, Malta, 26–27 April 2004*, 11–17.

Alsharaf, H., Cardey, S., Greenfield, P., Limame, D., Skouratov, I., 2003. Fixedness, the complexity and fragility of the phenomenon: some solutions for natural language processing. In *Proceedings of CIL17*, Hajičová, E., Kotěšovcová, A., Mírovský, J. (eds), CD-ROM.. Prague, Matfyzpress, MFF UK.

Alsharaf, H., Cardey, S., Greenfield, P., Shen, Y., 2004b, Problems and Solutions in Machine Translation Involving Arabic, Chinese and French. In *Proceedings of the International Conference on Information Technology, ITCC 2004, April 5–7, 2004, Las Vegas, Nevada, USA*,: 293–297. IEEE Computer Society.

Arnauld, A. & Lancelot, C. 1660. *Grammaire générale et raisonnée de Port-Royal*.

Bally, Charles. 1952. *Le language et la vie*. Genève: Droz.

Beauzée, Nicholas. 1767. *Grammaire générale ou, Exposition raisonnée des éléments nécessaires du langage, pour servir de fondement à l'étude de toutes les langues*. Paris: Barbou.

Benveniste, Emile. 1966. *Problèmes de linguistique générale*. Tome 1. Paris: Gallimard.

Birocheau, Gaëlle. 2000. Morphological Tagging to Resolve Morphological Ambiguities. In *Proceedings of the Second International Conference on Language Resources and Evaluation [LREC2000]*.

Birocheau, Gaëlle. 2004. Etiquetage morphologique et contribution à la désambiguïsation automatique des ambiguïtés morphologiques sur un lexique anglais. Thèse – PhD dissertation, Université de Franche-Comté, Besançon, France. Supervised by S. Cardey.

Bloomfield, Leonard. 1933. *Language*, New York: Holt, Rinehart & Winston.

Bopp, Franz F. 1833. *Vergleichende Grammatik des Sanskrit, Zend, Griechischen, Lateinishen und Deutschen*, Berlin: F. Dümmler.

Brøndal, Viggo. 1943. *Essais de linguistique générale*. Copenhagen: Munksgaard.

Buffier, Claude. 1709. *Grammaire françoise sur un plan nouveau pour en rendre les principes plus clairs et la pratique plus aisée*. Paris: N. Le Clerc et al.

Cardey, S., Greenfield, P., Wu, X. (2004), Designing a Controlled Language for the Machine Translation of Medical Protocols: The Case of English to Chinese. In *Proceedings of AMTA-2004 The 6th Conference of the Association for Machine Translation in the Americas, Georgetown University, Washington DC, USA, September 28 – October 2, 2004*, [Machine Translation: From Real Users to Research LNAI 3265] R.E. Frederking and K.B. Taylor (Eds), 37–47. Berlin, Heidelberg: Springer-Verlag

Cardey S. & Greenfield P. 2005. A Core Model of Systemic Linguistic Analysis. In *Proceedings of RANLP–2005* G. Angelova, K. Bontcheva, R. Mitkov, N. Nicolov (eds), 134–138.

Cardey, S., Anantalapochai, R., Beddar, M., Cornally, T., Devitre, D., Greenfield, P., Jin, G., Mikati, Z., Renahy, J., Kampeera, W., Melian, C., Spaggiari, L. & Vuitton, D. 2010. Le projet LiSe, linguistique, normes, traitement automatique des langues et sécurité: Du data et sense mining aux langues controlées. In *Proceedings WISG 2010, Workshop Interdisciplinaire sur la Sécurité Globale, Université de Technologie de Troyes, 26 & 27 January 2010*, CD-ROM.

Cardey S., Chan R., Greenfield P. 2006. The Development of a Multilingual Collocation Dictionary. In *Proceedings of COLING ACL 2006: MLRI 2006 workshop, Sydney, Australia, 23 July 2006* 32–40.

Cardey S., El Harouchy Z., Greenfield P. 1997. La forme des mots nous renseigne-t-elle sur leur nature? In *Actes des 5èmes journées scientifiques, Réseau Lexicologie, Terminologie, Traduction, LA MEMOIRE DES MOTS, Tunis*, Collection 'actualité scientifique' de l'AUPELF-UREF, 305–313.

Cardey S., Greenfield P. 2000. Génération automatique de déterminants et d'informations sur les déterminants en allemand en context. In *Actes du Colloque DET2000 – Détermination et Formalisation, Bellaterra, Université Autonome de Barcelone, 25 et 26 février 2000*, [Lingvisticæ Investigationes Supplementa 23]: 115–127.

Cardey S., Greenfield P. 2008. Micro-systemic linguistic analysis and software engineering: a synthesis. In *Revue RML6, Actes du Colloque International en Traductologie et TAL, 7 et 8 juin 2008, Oran*. F. Bouhadiba (ed), 5–25. Oran: Editions Dar El Gharb.

Cardey S., Greenfield P., Hong M-S. 2003. The TACT machine translation system: problems and solutions for the pair Korean – French. *Translation Quarterly* [27], 22–44. Hong Kong: The Hong Kong Translation Society.

Cardey, S, Devitre, D. Greenfield, P. Spaggiari, L. 2009. Recognising Acronyms in the Context of Safety Critical Technical Documentation. *In Proceedings ISMTCL*, [BULAG] S. Cardey (ed), 56–61. Besançon: PUFC.

Cardey, S. & Greenfield, P. 1992. The 'Studygram' Natural Language Morphology System: A First Step to an Intelligent Tutoring Platform for Natural Language Morphology. In *Proceedings of the UMIST ICALL workshop*. J. Thompson & C. Zähner (eds), 42–59. Hull: CTI Centre for Modern Languages, University of Hull.

Cardey, S. & Greenfield, P. 1997. Ambiguïté et traitement automatique des langues. Que peut faire l'ordinateur?. In *Actes du 16ème Congrès International des Linguistes, Paris, 20–25 juillet 1997*, Bernard Caron(ed), Elsevier Sciences.

Cardey, S. & Greenfield, P. 2003. Disambiguating and Tagging Using Systemic Grammar. In *Proceedings of the 8th International Symposium on Social Communication, Santiago de Cuba, January 20–24, 2003*. E. M. Bermúdes & L. R. Miyares (eds), 559–564. Santiago de Cuba: Centro de Lingüística Aplicada Ministerio de Ciencia, Tecnología y Medio Ambiente.

Cardey, S. 2008a. How to avoid interferences with other languages when constructing a spoken controlled language. In *Atti del congresso internazionale La comunicazione parlata Napoli, 23–25 febbraio 2006*: M. Pettorino, A. Giannini, M. Vallone, R. Savy (eds), Napoli: Liguori Editore.

Cardey, S. 2008b. Les langues contrôlées, fondements, besoins et applications. In *Proceedings WISG 2008, Workshop Interdisciplinaire sur la Sécurité Globale, Université de Technologie de Troyes, 29 & 30 January 2008*, 10 pages, CD-ROM.

Cardey, S. et al. 2005. (Haytham Alsharaf (Kuwait), Farouk Bouhadiba (Algeria), Sombat Khruathong (Thailand), Hsiang-I Lin (Taiwan), Xiaohong Wu (China), Kyoko Kuroda (Japan), Igor Skouratov (Russia), Valentine Grosjean (France), Gabriel Sekunda (Poland),

Izabella Thomas (Poland), Aleksandra Dziadkiewicz, (Poland), Duygu Can (Turkey)). Langues et cultures, systèmes et traduction, In *Journal des traducteurs* [vol 50, n°4 december 2005], revue META, CD ROM, Montréal.

Cardey, S. et al. 2006. (Greenfield P., Bioud, M., Dziadkiewicz H., Kuroda K., Marcelino I., Melian C., Morgadinho H., Robardet G., Vienney S.). The Classificatim Sense-Mining System. In *5th International Conference on NLP, FinTAL 2006 Turku, Finland, August 23–25, 2006 Proceedings* [Advances in Natural Language Processing LNAI 4139], T. Salakoski, F. Ginter, S. Pyysalo, T. Pahikkala (eds.), 674–684. Berlin, Heidelberg: Springer-Verlag.

Cardey, Sylviane et al. 2010. Le projet LiSe, Linguistique, normes, traitement automatique des langues et sécurité: du data et sense mining aux langues controlées. In *Proceedings WISG 2010, Workshop Interdisciplinaire sur la Sécurité Globale, Université de Technologie de Troyes, 26 & 27 January 2010*, 10 pages, CD-ROM.

Cardey, Sylviane. 1987. Traitement algorithmique de la grammaire normative du français pour une utilisation automatique et didactique, Thèse de Doctorat d'Etat, State Doctorat dissertation, Université de Franche-Comté, France, June 1987.

Cardey, S., Greenfield, P., Anantalapochai, R., Beddar, M., Devitre, D., Jin, G. 2008. Modelling of Multiple Target Machine Translation of Controlled Languages Based on Language Norms and Divergences. In *Proceedings of ISUC2008, Osaka, Japan, December 15-16, 2008*, B. Werner (ed), 322–329. IEEE Computer Society.

Carroll, Lewis. 1872. *Through the Looking Glass*. In: Alice's Adventures in Wonderland and Through the Looking Glass. Harmondsworth: Puffin Books, Penguin Books Ltd. (1974).

Chomsky, Noam. 1957. *Syntactic structures*. The Hague: Mouton.

Condillac, abbé Etienne-Bonnot de. 1771. *Traité des systèmes*. Paris: Arkstée & Merkus.

Descartes, René. 1637. *Discours de la method*. Leiden: De l'Imprimerie de Ian Maire.

Dial, R.,B. 1970. Algorithm 394. Decision Table Translation. *Communications of the ACM*, Vol. 12, No. 9, September 1970: 571–572.

Diderot, Denis et d'Alembert, Jean le Rond. 1751–1772. *Encyclopédie, ou Dictionnaire raisonné des sciences, des arts et des métiers*. Paris: Briasson, David, Le Breton & Durand.

Dubois, Jean. 1965. *Grammaire structurale du français*, tome 1, Paris: Larousse.

Ducrot Oswald, Todorov Tzvetan. 1968. *Dictionnaire encyclopédique des sciences du langage*. Paris: Seuil.

Durand, Daniel. 1987. *La systémique*. Paris: PUF.

Echchourafi, Hadnane. 2006. Vers une reconnaissance des composés pour une désambiguïsation automatique (composés à trois, quatre, cinq et six elements). Thèse – PhD dissertation, Université de Franche-Comté, Besançon, France. Supervised by S. Cardey.

Frei, Henri. 1929. *La grammaire des fautes*. Paris: Geuthner.

Gavieiro-Villatte, E. 2001. Vers un modèle d'élaboration de la terminologie d'une langue contrôlée; application aux textes d'alarmes en aéronautique pour les futurs postes de pilotage, Thèse de doctorat, direction Sylviane Cardey, Centre Tesnière, Université de Franche-Comté, France.

Gavieiro-Villatte, E., Spaggiari, L. 1999. Open-ended overview of controlled languages. In *Génie linguistique et genie logiciel*, [BULAG N° 24], P. Greenfield (ed), 89–101. Besançon: PUFC.

Gentilhomme, Yves. 1974. Les malheurs de Zizi. In *Le petit archimède n°8*: 170.

Gentilhomme, Yves. 1985. *Essai d'approche systémique, Théorie et pratique. Application dans le domaine des sciences du langage*. Berne, Francfort/main, New York: Peter Lang.

Girard, Abbé Gabriel. 1747. *Les vrais principes de la lanque française, VIIIe discours*, Paris: Le Breton

Gladkij, A.-V. & Mel'tchuk I.-A. 1969. *Elementy matematicheskoj lingvistiki*. Moscou: Ed. Nauka. 1972. *Eléments de linguistique mathématique*, Paris: Dunot.

Grand Larousse de la langue française. 1977. Paris: Larousse.

Greenberg, J.H. 1966. *Universals of language*. Cambridge, Mass. The Hague: Mouton & Co.

Greenfield P. 1998. Invariants in multilingual terminological dictionaries. In *Figement et traitement automatique des langues naturelles* [BULAG N° 23], P.-A. Buvet (ed), 111–121. Besançon: PUFC.

Greenfield, P. 1997. Exploiting the Model Theory and Proof Theory of Propositional Logic in the Microsystem Approach to Natural Language Processing. In *T.A.L. et Sciences Cognitives* [BULAG N° 22], H. Madec (ed), 325–346. Besançon: PUFC.

Greenfield, P. 2003. An initial study concerning the basis for the computational modelling of Systemic Grammar. In *Modélisation, systémique, traductabilité* [BULAG N° 28], S. Cardey (ed), 83–95. Besançon: PUFC.

Greimas, Algirdas, Julien. 1986. *Sémantique structural*. Paris: Presses universitaires

Grevisse, Maurice. 1980. *Le bon usage*. Paris: Gembloux, Duculot.

Guilbert, Louis. 1967. *Le vocabulaire de l'astronautique*, Université Rouen Havre.

Guillaume, Gustave. 1973. *Langage et sciences du langage*. Paris: Libr. A. G. Nizet.

Hagège, Claude. 1985. *L'homme de parole*. Paris: Payard.

Hagège, Claude. 1986. *La structure des langues*. Paris: PUF.

Harris, James. 1751. *Hermès, livre III, ch3*. London: H. Woodfall.

Harris, Zellig S. 1970. *Papers in structural and transformational linguistics*. Dordrecht-Holland, D. Reidel Publishing Company.

Hjelmslev, Louis. 1968. *Prolégomènes à une théorie du langage*. Paris: Minuit.

Hopcroft, J.E., Ullman, J.D. 1969. *Formal languages and their relation to automata*, Addison-Wesley Publishing Company, 1969.

Humboldt, Wilhem von. 1836. *Über die Kawisprache auf der Insel Jawa, Gesammelte Schriften*. Berlin: Druckerei der Königlichen akademie der wissenschaften.

Humby, E. 1973. *Programs from decision tables*. London/New York: Macdonald/American Elsevier

Jin, Gan & Khatseyeva, Natallia. 2012. A Reliable Communication System to Maximise the Communication Quality. In *8th International Conference on NLP, JapTAL 2012, Kanazawa, Japan, October 22–24, 2012. Proceedings*, [Advances in Natural Language Processing LNAI 7614], H. Isahara, K. Kanzaki (eds), Berlin, Heidelberg: Springer-Verlag

Jin, Gan. 2011. Chinese Sense-Mining for Intelligence and Security Domains. In *Proceedings of the 12th International Symposium on Social Communication – Comunicación Social en el Siglo XXI*, Santiago de Cuba, Cuba, January 17–21, 2011, Vol. I. E. M. Bermúdes & L. R. Miyares (eds), 163–166. Santiago de Cuba: Centro de Lingüística Aplicada Ministerio de Ciencia, Tecnología y Medio Ambiente.

Kiattibutra Anantalapochai, Raksi. 2011. Analyse lexicale morphologique et syntaxique du thaï en vue de la traduction automatique appliquée au domaine de l'administration publique. Thèse – PhD dissertation, Université de Franche-Comté, Besançon, France. Supervised by S. Cardey.

Knuth, D. 1975. *The art of computer programming*, Second edition, Volume 1 Fundamental Algorithms. Reading, Massachusetts: Addison-Wesley Publishing Company.

Kuroda, K., Chao, H.-L. 2005. *Divergence dans la traduction entre les langues orientales et le français*, [BULAG n° 30], K. Kuroda, H.-L. Chao (eds), Besançon: PUFC.

Landa, Lev Nahmandovič. 1965. Algorithmes et enseignement programmé. Grammaire de l'enseignement et type d'activité de penséé, 1971. In documents pédagogiques (textes traduits), n°4, ENS De Saint Cloud, Centre de recherche et de Formalisation en Education, 1976.

Le Robert electronic dictionary of French. 1997.

Lopota, G. 1981. Bibliothèque de Q.C.M,- It 603 – (A04,A23 ...) Resumé des principales notions utilisées sur les relations binaires.

Martinet, André. 1973. *Eléments de linguistique générale*. Paris: Colin.

Mikati, Ziad. 2009. Data and Sense Mining and their Application to Emergencies and to Safety Critical Domains. In *Proceedings ISMTCL*, [BULAG] S. Cardey (ed), 179–184. Besançon: PUFC.

Mohd Nor Azan Bin Abdullah. 2012. The Malay verb system. In *Proceedings of The 6th International Workshop on Malay and Indonesian Language Engineering, MALINDO 2012*, Bali Ranaivo-Malançon (ed), 1-6. Universiti Malaysia Sarawak (UNIMAS), Kuching, Sarawak, Malaysia: MASAUM Network (CDROM).

Mounin, Georges. 1974. *Dictionnaire de la linguistique*. Paris: PUF

Palmer, F. 1975. *Grammar*. Harmondsworth: Penguin Books.

Panini, 1966. *La grammaire de Panini*. Traduction Louis Renou. Ecole française d'extrème-orient. Paris, Hindi Press Private Ltd.

Pascal, B. 1670. *Pensées de Pascal dans Œuvres de Blaise Pascal*, œuvre posthume. English translation: *Pascal's Pensées*. 1958. New York NY: E.P. Dutton & Co.

Pott, August-Friedrich. 1849. Jahrbuch der Freien Deutschen Akademie. Frankfurt.

Pottier, Bernard. 1962. *Systémique des éléments de relation*. Paris: Klincksieck

Routledge Dictionary of Language and Linguistics. 1996. Bussmann Hadumod translated and edited by Trauth Gregory P., Kerstin Kazzazi, London, New York: Routledge.

Sanctius, Franciscus. 1587. *Minerva sive de causis linguae latinae*. Salamanca: Renaut.

Sapir, E. 1921. *Language: An Introduction to the Study of Speech*. New York NY: Harcourt, Brace and Company.

Saussure, Ferdinand de. 1922. *Cours de linguistique générale. Course of General Linguistics*, translated and annotated by Roy Harris, 1990. London: Duckworth.

Scaliger, Julius Caesar. 1540. *De causis linguae latinae*. Ed. Lyon.

Schlegel, August Wilhelm von. 1818. *Observation sur la langue et la littérature provençales*. Paris: Librairie grecque-latine-allemande.

Schlegel, Friedrich von. 1808. *Über die Sprache und Weisheit der Indier*. Heidelberg: Mohr & Zimmer.

Schleicher, August.1866. *Compendium der vergleichenden Grammatik der indogermanischen Sprachen*. Weimar: Böhlau.

Shuxiang. 1979. 现代汉语八百词 Xiandai hanyu babai ci, 吕叔湘Lü Shuxiang, P175, ISBN7100013097. ("现代汉语八百词Xiandai hanyu babai ci" is the book title, "吕叔湘Lü Shuxiang" is the author).

Spaggiari, L., Beaujard, F., Cannesson, E. 2005. A controlled language at Airbus. In *Machine Translation, Controlled Languages and Specialised Languages*, [Lingvisticæ Investigationes, tome XXVIII], S. Cardey, P. Greenfield, S. Vienney (eds), 107–122. Amsterdam: John Benjamins.

Spivey, J.M., *The Z Notation*, Hemel Hempstead: Prentice Hall,1992.

Steinthal, Heymann. 1860 *Charakteristik der hauptsächlichsten typen des sprachbaues*. Berlin: F. Dümmler

Tesnière, Lucien.1982. *Eléments de syntaxe structurale*. Paris: Klincksieck.

Trakhtenbrot, B. A. 1960. *Algorithms and Automatic Computing Machines.* Translated and adapted from the second Russian edition (1960). Lexington, Massachusetts: D. C. Heath and Company, 1963.

Trevoux. 1734. *Dictionnaire dit de Trevoux.* Paris.

Troubetzkoy, Nicolas, Sergueevitch. 1976. *Principes de phonologie.* Trad. by J. Cantineau. Paris: Klincksieck

Turing A. M. 1936. On Computable Numbers, with an Application to the Entscheidungsproblem, *Proceedings of the London Mathematical Society, Series 2, Vol.42 (1936 – 37)*: 230–265, with corrections *Proceedings of the London Mathematical Society, Series 2, Vol.43 (1937)* 544–546.

Vauquois, Bernard.1975. *La traduction automatique à Grenoble.* Documents de linguistique quantitative. Paris: Dunod.

Vienney, S., Cardey, S., Greenfield, P. 2004. Systemic analysis applied to problem solving: the case of the past participle in French. In *4th International Conference on NLP, EsTAL 2004, Alicante, Spain, October 20–22, 2004, Proceedings* [Advances in Natural Language Processing LNAI 3230], J. L. Vicedo, P. Martiínez-Barco, R. Muñoz, M. Saiz Noeda (eds), 431–441. Berlin, Heidelberg: Springer-Verlag.

Vives, Joannes Ludovicus. 1531. *De Causis Corruptarum Artium.* De Disciplinis Libri XX.

Wagner, Robert-Leon.1973. La grammaire française, Paris: S.E.D.E.S.

Whitney, William Dwight. 1867. *Language and the study of language, Twelve Lectures on the principles of Linguistic Sciences,* London: Trübner.

Whitney, William Dwight. 1875. *La vie du langage.* Paris: Librairie Germer-Baillière.

Zaretskaya, Anna. 2012. Automatic E-mail Answering Based on Text Classification: a Rule-based Approach, Masters dissertation (Erasmus Mundus International Masters in Natural Language Processing and Human Language Technology), Université de Franche-Comté.

Index